Recent Progress in Organic Farming

Recent Progress in Organic Farming

Editor: Cruz Hawkins

R CALLISTO REFERENCE

www.callistoreference.com

Callisto Reference,
118-35 Queens Blvd., Suite 400,
Forest Hills, NY 11375, USA

Visit us on the World Wide Web at:
www.callistoreference.com

ISBN: 978-1-64116-190-9 (Hardback)

Cataloging-in-Publication Data

Recent progress in organic farming / edited by Cruz Hawkins.
      p. cm.
Includes bibliographical references and index.
ISBN 978-1-64116-190-9
1. Organic farming. 2. Agriculture. I. Hawkins, Cruz.
S605.5 .R43 2019
631.584--dc23

Table of Contents

Preface

Organic farming is an agricultural practice, characterized by the use of products of organic origin like compost manure, bone meal, green manure, etc. It also involves the techniques of companion planting and crop rotation. Organic farming is a system that discourages the use of plant growth regulators, antibiotics in livestock, hormones, genetically modified organisms, etc. It strives to adopt a healthy, sustainable and self-sufficient approach to farming. Crop diversity, composting, biological pest control, integrated pest management, etc. are some of the common practices of this field. The study of organic farming methods is under the scope of agroecology. This book unfolds the innovative aspects of organic farming which will be crucial for the progress of this field in the future. Also included in this book is a detailed explanation of the various concepts and practices of this discipline. As this field is emerging at a rapid pace, the contents of this book will help the readers understand the modern concepts of the subject.

Significant researches are present in this book. Intensive efforts have been employed by authors to make this book an outstanding discourse. This book contains the enlightening chapters which have been written on the basis of significant researches done by the experts.

Finally, I would also like to thank all the members involved in this book for being a team and meeting all the deadlines for the submission of their respective works. I would also like to thank my friends and family for being supportive in my efforts.

Editor

Role of Organic Sources of Nutrients in Rice (*Oryza sativa*) Based on High Value Cropping Sequence

Sanjay Kumar Yadav, Subhash Babu, Gulab Singh Yadav, Raghavendra Singh and Manoj Kumar Yadav

Additional information is available at the end of the chapter

Abstract

The organic nitrogen (N) nutrition of organic manuring with biofertilizers had the highest rice equivalent grain yield, production efficiency, net energy return, as well as net monetary return and profitability in rice-based cropping sequence. The different rice-based cropping sequences did not differ with respect to yield and quality parameters. However, the organic N nutrition with organic manures along with biofertilizers proved significantly superior with respect to yield and quality parameters of rice, potato, and onion, respectively. The different rice-based cropping sequences differ with respect to nutrient uptake, e.g., rice-maize-onion had the highest removal of major (N, P, K), secondary (S), and micronutrients (Zn, Fe, Mn, Cu) than the rest of cropping sequence, which was significantly superior to the rest of the sequences. The organic N nutrition with organic manures along with biofertilizers proved superior due to its visible favorable effect on soil health with respect to nutrient status and microbial count and this indicates the utilization of this low-cost but long-term beneficial practice under high-intensity cropping for sustainable crop production.

Keywords: Biofertilizers, organic farming, high value crops, cropping sequence

1. Introduction

Organic farming is a production system that avoids or largely excludes the use of synthetic fertilizers, pesticides, growth regulators, and livestock feed additives. The objectives of environmental, social, and economic sustainability are the basics of organic farming.

The maintenance of good soil fertility is essential for sustainable crop production, which requires the regular use of organic sources of nutrient-like organic manure and biofertilizers to keep the farm income higher of the farming community. Organic agriculture is a holistic production management system, which promotes sustainable agriculture and enhances agro ecosystem health, including biodiversity, biological cycle, and soil biological activity. The organic farming practices on scientific principles are as productive as the conventional system. Organic systems showed greater soil health benefits reduced cost on production, are found better than inorganic practices, and enhanced profit margin with quality food. Interestingly, while exports of organic commodity are growing, domestic market demand is galloping for high-value crop produce, supports from government are increasing and innovation system support has started to grow. In such situation, it is necessary to develop suitable technology for meeting the challenges of the coming generation by providing good quality produce without deteriorating the socio-economic conditions of the farmer and with minimum environmental pollution. The farmers of ancient India adhered to the natural laws and this helped in maintaining the soil fertility over a relatively longer period of time [1]. These organic sources, besides supplying N, P, K, also make unavailable sources of elemental nitrogen, bound phosphates, micronutrients, and decomposed plant residues into available form in order to facilitate the plants to absorb the nutrients. Organic cultivation practices are very effective to improve the population of beneficial microorganisms in the soil having direct effect on enhancing the availability of macronutrients and micronutrients through correcting the deficiency induced by the conventional practices with the application of synthetic fertilizers, and consequently capable of sustaining high crop productivity and soil biological properties by modification of the soil environment [2].

The farmers can in turn, get good remuneration from the organically produced crops and vegetables if included in high-value crop sequences, e.g., aromatic rice–table pea and onion [3] due to their heavy demands in domestic, national, as well as international markets that may help the country in earning some foreign exchange. Therefore, a book chapter entitled "Role of organic sources of nutrient in rice (*Oryza sativa*) based on high value cropping sequence" was planned and executed with the following objectives:

1. To identify potential high-value cropping sequence suitable for irrigated ecosystem;

2. To study the effect of organic nitrogen sources on yield and quality of crop produce;

3. To study the effect of organic nitrogen sources on nutrient acquisition by the sequence.

2. Experimental details

2.1. Treatment details

2.1.1. Main plot: Cropping sequences (7)

- Sequence-1: Rice-Potato-Onion

- Sequence-2: Rice-Green Pea-Onion

- Sequence-3: Rice-Potato-Cowpea (Green Pod)

- Sequence-4: Rice- Green Pea -Cowpea (Green Pod)

- Sequence-5: Rice-Rajmash (Green Pod)-Onion
- Sequence-6: Rice-Rajmash (Green Pod)-Cowpea (Green Pod)
- Sequence-7: Rice-Maize (Green Cob)-Cowpea (Vegetable)

2.1.2. Sub plot: Manurial treatments (3)

- Control (without organic manures)
- 100% RDN through organic manures as 1/3 FYM + 1/3 Poultry Manure (PM) + 1/3 Vermi-compost
- 100% RDN through organic manures as 1/3 FYM + 1/3 Poultry Manure (PM) + 1/3 Vermi-compost + Azotobacter + PSB

Crop	Variety	Seed rate (kg ha⁻¹)	Spacing (cm)
		Rainy season	
Rice	HUBR 2-1	40 kg	20 x 15
		Winter season	
Maize (Green Cob)	Pioneer Hybrid	20 kg	60 x 20
Green Pea	Early Apoorva	80 kg	30 x 10
Rajmash	HUR-137	80 kg	30 x 10
Potato	Kufri Badshah	2,000 kg	50 x 25
		Summer season	
Onion	Agrifound Light Red	10 kg	20 x 15
Cowpea	Tokito Hybrid	10 kg	50 x 20

Table 1. Details of the variety of hybrid seed rate and spacing of different crops.

3. Rainy season (rice)

3.1. Field preparations

Proper field preparation and timely planting are essential for good crop yield. These factors influence the soil's physical property, particularly soil moisture, aeration, and plant nutrient availability. With a view to have good experimental unit for planting, initial ploughing was done by a soil turning plough followed by disking. The seed beds were properly prepared as per crop requirements before planting various crops.

3.2. Raising rice nursery

A well-drained fertile land having good irrigation facility was selected for raising rice seedlings. The nursery plot was ploughed twice and puddled in standing water to convert the upper layer of soil into fine soft mud. The field was leveled properly and 10 x 1.5 m² beds were prepared. A requisite amount of 36 kg organic manure was applied to each nursery of 15 m². Healthy, genuine, certified, and sprouted seeds at 40 kg per ha were properly spread, keeping a thin water film for a week. The seedbed was irrigated to maintain shallow, submerged rice.

3.3. Field preparation for transplanting

Proper field preparation is essential for a healthy rice crop. The experimental area was ploughed with a tractor during the summer and ploughed twice again before rice transplanting. Thereafter, the field was puddled with the cultivator. Finally, the field was laid out to meet the requirements of the experimental design. The field was puddled thoroughly, and four-week-old seedlings were transplanted at 3 seedlings per hill in rows 20 cm apart with hill to hill distance of 10 cm. As per treatment, full recommended doses of all the manures were applied just before transplanting. Irrigations were given to the crop at 16, 30, 18, and 32 DAT during the two years of experimentation. Two hand weedings were done at 26 and 65 DAT during both the years of experimentation. Except minor appearances of gundhi bugs, no major pests or diseases appeared. Hence, even bio-insecticides were not used due to the negligible impact of the gundhi bugs. Rice plants were harvested at physiological maturity of the crop after 108 DAT during the first and 109 DAT in the second year of experimentation. First of all, the border rows were harvested, bundled, and removed from the plots. Thereafter, the experimental rows from the net plot area were harvested. Plot wise harvested materials were carefully bundled, tagged, and taken to the threshing floor. Each bundle was weighed after complete sun drying and threshing. The grain yield was recorded separately after winnowing and cleaning. The straw yields was calculated by subtracting grain yield from the bundle weight and were converted to kg per ha based on net plot size harvest.

3.4. Biometric observations of rice

For recording biometric observations at different stages of crop growth, four hills in the net plot area were randomly selected and tagged. However, for the dry matter production, four hills were randomly selected from the sample rows (border rows) at different growth stages. The plants were then tagged and brought to the laboratory for the study. Four biometric observations were recorded at 30 DAT (tillering stage), 60 DAT (late jointing stage) and at harvest during both years. The plant samples collected randomly from the border row of the field were kept in an oven at 60°C till the constant weight arrived for determining the dry matter production per unit area. The panicle-bearing tillers were counted from the one square meter marked area after full anthesis. Ten panicles were randomly selected from tagged plants and the length was measured from the neck node to the tip of the upper most spikelet and average length was recorded. Ten randomly selected panicles were weighed and averaged to record per panicle weight. The filled grain of each of the ten panicles from each plot were counted and averaged. Grain samples were taken from the threshed and cleaned produce of each net plot and 1,000 grains were counted and weighed. Grain yield was recorded (kg plot⁻¹) after threshing, winnowing, cleaning, and drying. Thereafter, it was computed to kg per ha.

The difference of the bundle weight and grain yield gave the straw yield (kg plot^{-1}). Thereafter, it was computed to kg per ha.

4. Winter season

4.1. Field preparation

During the winter season, potato, green pea, rajmash, and maize are grown. The following packages of practices were adopted for these crops. Field preparation operations were common for all the *rabi* season crops. As a general rule, these crops require a well pulverized but compact seedbed for good and uniform germination. To avoid the mixing of soil under treatments, the individual plot was ploughed thrice by power tiller at proper tilth and finally the planking was done.

4.2. Weed management

During both years of experimentation, the weeding was done using a hand rotary weeder during the beginning of the first appearance of a thick flush of weed, e.g., 25 days after sowing followed by a second weeding at 45–50 days after sowing. The first weeding was done after recording observations for weed flora. However, to the wheat crop, only one weeding was given.

4.3. Irrigation

In both years of the experiment, irrigation was given according to the requirements of the different crops as per the schedule. In all, one irrigation was given to lentil, pea, and chickpea, two irrigations to mustard, three irrigations to potato and wheat, and as much as four irrigations was given to maize. Only minor appearances of pests or diseases occurred. Hence, even bio-insecticides were not used due to the negligible impact of the insect pests and diseases.

4.4. Harvesting

In general, all the crops were harvested by serrated edge sickle manually at the maturity of the respective crops. However, in case of potatoes, tubers were dug out at maturity. In green peas, two to three pickings of green pods were done; whereas, the green cobs of maize were harvested at the milky stage of the grains. Haulms of pea and maize stover were used as cattle fodder. In all the crops, the border rows and 0.5 m either side of plot rows were harvested and removed around the individual plots leaving only the net plot area. The harvesting of each net plot area was done separately and the harvested material from each plot was carefully bundled, tagged, and taken to the threshing floor and kept individually for sun drying.

4.5. Threshing

Each bundle was weighed after proper sun drying and then threshed individually. The grain/seed/pod/tuber yield of different crops were weighed and recorded separately after winnowing and cleaning. The straw stover yield were calculated/recorded separately and converted to q ha^{-1} based on the net plot size harvest.

5. Summer season

5.1. Field preparation

Onions and cowpeas were taken during summer season in different cropping sequences. Field preparatory operations were common for all summer season crops. After the harvesting of winter season crops in different sequences, pre-sown irrigation was given and individual plots were tilled thrice with a power tiller at proper tilth and finally planking was done.

5.2. Raising of onion seedling

Seeds of Agrifound light red variety were used. The seeds used for the nursery had more than 80% germination. The nursery beds (4 m x 2.6 m) were prepared carefully by incorporating sufficient quantity of well-rotten farm yard manure (20 kg bed^{-1}). Seeds were sown on the bed at 52 g per bed. After sowing, beds were given light and frequent water application through a water cane at the beginning to maintain moisture for seedling growth. Two light irrigations were also given at sowing and 10 DAS to maintain the growth of a thin layer of FYM was given to cover the seeds. The beds were covered with a thin layer of paddy straw on the same day to maintain congenial moisture and temperature condition. The paddy straw was removed after seed germination (10 DAS). Seedlings were transplanted at 60 DAS on 26.02.04 during the first year and 20.02.05 during the second year. However, cowpea seeds were treated with *Rhizobium* culture to improve the nitrogen fixation capacity before sowing the crop. Details of crop varieties used, seed rate, and spacing are given in Table 1.

5.3. Weed management

During both years of the experimentation, one weeding was done in the inter-row spaces by hand rotary weeder at 20 days after sowing and the weeds on the crop rows were removed manually.

6. Qualitative character of rice-based cropping sequence

6.1. Hulling of rice (%)

Two hundred grams of rice grains after threshing, winnowing, cleaning, and drying were taken for dehusking, and the brown rice thus obtained was weighed and then hulling (%) was calculated by the following formula:

$$\text{Hulling}(\%) = \frac{\text{Brown rice obtained after threshing (g)}}{\text{Total rice grain taken for dehusking (g)}} \times 100$$

6.2. Milling of rice (%)

One hundred grams of brown rice obtained after hulling was taken and kept for polishing by removing rice bran, embryo, and alurone layer and polished white kernels were thus obtained using the following formula:

$$\text{Milling}(\%) = \frac{\text{White polished kernels obtained (g)}}{\text{Brown rice taken for polished (g)}} \times 100$$

6.3. Head rice recovery (%)

Total white polished rice obtained after milling was taken and whole white kernels were separated, weighed and the percentage was calculated using the formula:

$$\text{Head Rice Recovery}(\%) = \frac{\text{Whole white kernels obtained (g)}}{\text{White polished kernels obtained after milling (g)}} \times 100$$

6.4. Shelling of maize (%)

Five randomly selected cobs were weighed and grains were separated and weighed. The shelling percentage was calculated by using the following formula:

$$\text{Shelling}(\%) = \frac{\text{Weight of kernels per cob (g)}}{\text{Weight of cob (g)}} \times 100$$

6.5. Protein content of each crop (%)

The protein content (%) in the grains was worked by multiplying the nitrogen content in grain by the factor 6.25 (A. O. A. C., 1960).

6.6. Protein yield

The protein yield (kg ha^{-1}) was obtained by the following formula:

$$\text{Protein Yield}(\text{kg/ha}) = \frac{\text{Protein content per cent} \times \text{Yield (kg/ha)}}{100}$$

6.7. Starch content of potato (%)

It was extracted and determined according to Carillo et al (2005).

Pungency estimation of allyl-propyl-disulphide onion: Allyl-propyl-disulphide content in the onion bulb was determined as Pyruvic acid (Hort and Fisher, 1970) using the following relationships:

$$\text{Pyruvate content} = \frac{\text{Pyruvate content from standard curve } (\mu \text{ mo})}{\text{Alliquat of test control solution taken color development } (\text{ml})} \times \frac{\text{Total volume of soln. of sample made } (\text{ml})}{\text{Wt of sample taken for assay } (\text{g})}$$

Carbohydrate content (%): It was determined by the method described by Loomis and Shull (1937).

Nutrient content: The seed and plant samples at harvest were used for chemical analysis of N, P, and K contents. The plants and seeds were dried in an oven and grained thoroughly in a wily mill to pass through a 30-mesh sieve. These were presented in labeled polythene bag for chemical analysis.

Total nitrogen: The total nitrogen was analyzed at harvest. The N content in seeds was also analyzed separately. Alkaline permanganate method [4] was employed for their estimation.

Total phosphorous: Total phosphorus was estimated during the harvest of the crop 0.05 M $NaHCO_3$ using Barten's reagent was employed for this purpose.

Total potassium: Total potassium was determined with the help of a flame photometer [5] during the harvest of both seed and straw.

Micronutrient: Micronutrient was determined with the help of an atomic absorption sector-photometer at the time of harvesting of both seed and straw.

Nutrient uptake: Nutrient uptake in grain (seed/bulb) and straw/haulm of the crops were calculated in kg/ha in relation to yield by using the following formula:

$$\text{Nutrient Uptake (kg/ha)} = \frac{\text{Nutrient content (\%)} \times \text{Yield (kg/ha)}}{100}$$

7. System study

Equivalent yield: Rice equivalent as well as system productivity were worked out by converting the yields of crops into rice equivalent, taking the help of price values used for the calculation of the economics. The productivity of cropping sequence was converted into rice equivalent yield using the formula:

$$\text{Rice Eq.Yield (kg/ha)} = \frac{\text{Productivity of component (kg/ha)} \times \text{Price of component (Rs/kg)}}{\text{Cost of Rice (Rs/kg)}}$$

Equivalent yields of potato and onion were also calculated as same manner as fallow in calculating rice equivalent yield.

Production efficiency of the system (PES): Production efficiency of the system was calculated by dividing the equivalent yield of rice in a sequence through 365.

$$\text{PES (kg/ha/day)} = \frac{\text{Rice equivalent yield of the system (kg/ha) in a year}}{365}$$

Nutrient uptake in the system: Nutrient uptake in the system was worked out by making a sum of nutrient uptake of a sequence.

Economic analysis

a. Cost of cultivation: The cost of cultivation of various sequences was worked out based on the most recent standard rate of materials.

b. Gross return: The yield of different component crops in the sequence were converted into gross return in rupees based on the current market price.

c. Net return: Net return for each crop sequence was calculated by deducting the cost of cultivation from the gross return.

Cost of cultivation, gross return, and net return under different treatments were worked out on the basis of prevailing cost of different inputs. Power and labor for different operations were calculated on a per hectare basis as per normal rates prevalent in the country. The costs of other inputs were considered as per market price. The total gross return was taken as the total income received from the produce of economic and stover yield. Net return was calculated with the help of following formula:

Net Return = Gross Return - Cost of Cultivation

Energy equivalent: The total energy return of the system was obtained by the conversion of economic yield of the sequence into energy equivalent; whereas, the net energy return was worked out by deducting total input involved in the sequence in energy term from the total energy return. The energy output: input ratio and energy productivity were obtained as follows:

$$\text{Energy Output: Input Ratio} = \frac{\text{Total energy return}}{\text{Total input involved in terms of energy}}$$

The various practices involved in crop production and economic yield of component crops in the sequences were converted into the equivalent value of chemical energy (MJ/ha). For these conversions, standard values as given by [6] were used (Table 2).

S. No.	Particulars	Units	Equivalent energy (MJ/ha)
Input			
1.	Human labor		
	Adult men	Man hours	1.96
	Women	Woman hours	1.57
2.	Diesel	Liter	56.31
3.	Electricity	KWH	11.93
4.	Chemical fertilizer		
	(a) Nitrogen	Kg	60.6
	(b) P_2O_5	Kg	11.1
	(c) K_2O	Kg	6.7
5.	Plant protection (Superior)		
	Granulated chemical	Kg	120
	Liquid chemical	ml	0.102
6.	Seeds		
	Potato	Kg	4.06
	Rice, maize	Kg	14.7
	Onion	Kg	15.8
	Cowpea, pea, rajmash	Kg	14.7
Output			
1.	Rice	Kg (dry mass)	14.7
	Cowpea, table pea, rajmash	Kg (pod)	3.89
2.	Onion	Kg (bulb)	2.60
3.	Potato	Kg (tuber)	4.06
4.	Maize	Kg (green cob)	4.41

Table 2. Energy coefficients.

8. Effects of organic sources in rice based on cropping sequence

8.1. Effect of weather in crops

Plants growing in natural environment are often prevented from expressing their full genetic potential for production as they are subjected to various biotic and abiotic stresses. Environmental factors are relatively more dynamic in determining the extent of growth and development of plants and play major roles in the completion of the plant life cycle. Every crop requires a definite set of environmental conditions for its proper growth and development. Matching the crop phenology to the climatic environment prevailing during the growing season is an important aspect to maximize genetic yield potential.

8.2. Economic yield of rice

In organic nitrogen sources, the application of 100% RDN through organic manure along with biofertilizers recorded the highest grain yield during both years of investigation. This might be due to better availability of nutrients through superimposition of organic manure along with biofertilizers. It was also observed that plants were well supplied with nitrogen, senescence of flag leaf was delayed, and respiratory losses were low. Potassium also had expressed, in addition of CO_2 assimilation rates, resulting in more supply of photosynthates along with micronutrients responsible for the effective translocation of photosynthates that probably accounted for the highest economic yield. In addition to these, *Azotobacter* produced growth promoting substances that improved seed germination and growth with extended root systems. It also produced polysaccharides that improved soil aggregation; whereas, PSB in the rhizosphere of rice rendered insoluble soil phosphate available to plants due to their production and secretion of organic acids, as well as due to the release of sufficient amounts of nitrogen by mineralization at a constant level, which in turn resulted in better crop growth and improvement in various yield components of rice.

8.3. Potato equivalent yield of winter season crops

The maximum potato equivalent yield was recorded under the sequence rice-potato-cowpea (green pod). It may be emphasized here that PEY of crops is the function of market price along with the yield of a particular crop. The potato itself produced higher economic yield and this is accompanied with better market value as a result of potato equivalent yield that were higher as compared to other sequences. Further nitrogen application through organic manures significantly augmented the potato equivalent yield due to the continuous raising of organic potato bio-dynamically on the same site, which improved tuber production by enriching soil fertility.

8.4. Onion equivalent yield of summer season crops

The maximum onion equivalent yield was recorded under the sequence rice-green pea-onion. The onion itself produced higher economic yield due to the inclusion of legume as a previous crop and this accompanied with better market value as a result of onion equivalent yield that were higher compared to other sequences. Further nitrogen application through organic manures significantly augmented the onion equivalent yield, which was due favorable growth and yield of onion crop.

9. Effect on quality parameters

9.1. Rice

The application of organic nitrogen also influenced protein content and protein yield due to the increase in the concentration of nitrogen in grains, which might have modified the proportion of grain constituents. The higher uptake of nutrients, particularly nitrogen, in the organic nitrogen treatments was probably responsible for the higher grain protein. Accumulation of protein in seeds may also be increased due to the continuous nitrogen supply and its translocation in seed buds and optimal nutrition. It is known that protein content imparts strength to the grain; higher protein content thus resulted in higher head rice recovery.

9.2. Potato

Amongst various nitrogen substitution treatments, maximum starch content was recorded under organic sources of nitrogen along with biofertilizers, especially due to higher concentration of potassium in poultry manure, which might have modified the proportion of tubers constituents with respect to starch.

9.3. Onion

Application of organic nitrogen significantly increased the allyl-propyl-disulphide and carbohydrate content (%) in onion bulbs might be due to increased volatile fatty oil content resulting in significantly higher production of allyl-propyl-disulphide in onion bulbs. Increased allyl-propyl-disulphide content with increasing organic nitrogen application was in close agreement with findings of [7, 8].

10. System analysis

10.1. Rice Grain Equivalent Yield (RGEY)

The maximum RGEY was recorded under the sequence rice-potato-onion. The higher production potential of potato and onion and better market prices were instrumental for attaining higher REY by this sequence [9, 10]. Rice equivalent yield is directly associated with the yield of respective crops in the sequence and so organic manure alone or along with biofertilizers enhanced the yield potential of crops, which ultimately increased the rice equivalent yield of the sequence.

10.2. Production efficiency

The sequence rice-potato-onion had recorded maximum production efficiency compared to the rest of the treatments and this was due to the better market price of potato and onion in the sequence [11]. Organic manures along with bio-fertilizers recorded the significantly highest production efficiency of the system and this was due to the highest rice grain equivalent yield of crops in the system.

10.3. Energetics

The maximum energy input was recorded in the rice-potato-onion sequence. The energy consumed by the potato through fertilizer, seeds, and human labor and that of the onion for irrigation (electricity) and inter-culture operations resulted in higher energy input. The energy involved in N fertilizer was particularly higher in sequences involving potato and onion, which relatively consumed a large proportion of energy in seeds. The pooled data indicated that the maximum gross energy output, net energy return, and employment generation was obtained in the rice-potato-onion sequence. This clearly exhibited that besides having more energy input, this sequence also produced the highest energy equivalent, resulting into maximum gross energy output, net energy return, and employment generation [12]. In general, the gross energy output, net energy return, and employment generation of the system remained comparatively higher during the second year than that of the first year. Application of nitrogen through organic manures along with bio-fertilizers recorded maximum average energy input, gross energy output, net energy return, and employment generation of the system because this sequence was more input intensive as well as had the highest productivity level.

10.4. Economics

Data related to economics as affected by various cropping sequences and organic nitrogen treatments of two years of experimentation are presented. The maximum cost of cultivation, gross return, net return, and profitability was recorded under the sequence rice-potato-onion, which was significantly higher than that of the other sequences. This was mainly due to the higher production potential of potato, accompanied with good monetary return from the onion. The highest values of cost of cultivation, gross return, net return, and profitability were associated with the application of nitrogen through organic manures along with biofertilizers. This was mainly due to higher productivity without a proportionate increase in the cost of cultivation.

10.5. Nutrient uptake

Nutrient uptake by different cropping sequences is the function of crop yield and nutrient content. The increase in these factors was responsible for the increased nutrient uptake during both years of experimentation of the system, which was at the maximum under the rice-cowpea-maize sequence. This was significantly superior to the rest of the sequences in this respect, which could be a higher productivity potential of maize ascribed to the increase in the available nitrogen, phosphorus, potassium, sulfur, zinc, iron, copper, and manganese contents in the soil resulting from the increased availability of nutrients through organic sources particularly through organic manure along with biofertilizers.

10.6. Soil fertility status

Data on the nutrient status of soil organic carbon, major (nitrogen, phosphorus potassium), secondary (sulfur), and micronutrients (zinc, iron, copper, and manganese), recorded maximum improvement, in this respect, was observed where pulse crops were incorporated in the sequence. Application of either organic manure alone or with biofertilizers significantly improved the soil status with respect to organic carbon and nutrients under study. It is quite obvious that this might have added greater organic sources and biofertilizer to the soil, ultimately improving the soil's organic carbon. Similarly, [13] also reported that 100% nitrogen (1/3 each from cow dung manure, neem cake, and composed crop residue) appreciably increased the organic carbon (6.3 g kg^{-1}) over the initial value (5.8 g kg^{-1}).

10.7. Soil health

The application of organic manure along with biofertilizer significantly improved soil pH, as well as electrical conductivity was associated with the decline in soil reaction might be due to organic compounds added to the soil in the form of organic manure and biofertilizer that produced more humus and organic acids in decomposition. The role of organics is attributed to the supply of essential nutrients by the continuous mineralization of organic manures, nutrient supplying capacity of the soil, and its favorable effect in the soil's biological (bacteria, actinomycetes and fungi) properties [14,15]

11. Conclusion

1. The inclusion, of the two high-value vegetable crops in sequence having 300%, rice-potato-onion had the highest rice equivalent grain yield, production efficiency, net energy return, as well as net monetary return and profitability. However, the best benefit ratio was highest in the sequence rice-potato-cowpea (green pod). Thus, rice-potato-onion was observed as the most intensive, stable, and profitable high-value cropping sequence for irrigated ecosystems.

2. The organic N nutrition of organic manuring with biofertilizers had the highest rice equivalent grain yield, production efficiency, net energy return, as well as net monetary return and profitability on rice-based cropping sequence.

3. The different cropping sequences of rice did not differ with respect to yield and quality parameters. However, the organic N nutrition with organic manures along with biofertilizers proved significantly superior with respect to the yield and quality parameters of rice, potato, and onion.

4. The different cropping sequences of rice differ with respect to nutrient uptake, i.e., rice-maize-onion had the highest removal of major (N, P, K), secondary (S), and micronutrients

(Zn, Fe, Mn, Cu) than the rest of the cropping sequences and was significantly superior to rest of the sequences.

5. The organic N nutrition with organic manures along with biofertilizers had the highest nutrient acquisition of major (N, P, K), secondary (S), and micronutrients (Zn, Fe, Mn, Cu).

6. The different cropping sequences of rice did not differ with respect to nutrient status as well as microbial count. However, inclusion of pulses in sequences showed positive improvement on soil health and the effect can be quite effective and visible on a long term basis.

7. The organic N nutrition with organic manures along with biofertilizers proved superior due to its visible favorable effect on soil health with respect to nutrient status and microbial count and this indicates the utilization of this low-cost but long-term beneficial practice under high intensity cropping for sustainable crop production.

Author details

Sanjay Kumar Yadav*, Subhash Babu, Gulab Singh Yadav, Raghavendra Singh and Manoj Kumar Yadav

*Address all correspondence to: sanjaybhu05@rediffmail.com

ICAR-Central Potato Research Station, Upper Shillong, Meghalaya, India

References

[1] Chandra, S. and Chauhan, S.K. (2004). Prospects of organic farming in India. *Indian Fmg.* May 11-14.

[2] Yadav, S.K.; Babu, S.; Yadav, M.K.; Singh, K.; Yadav, G.S.; and Pal Suresh. (2013). A Review of Organic Farming for Sustainable Agriculture in Northern India, *International Journal of Agronomy*, vol. 2013, Article ID 718145, 8 pages, 2013. doi: 10.1155/2013/718145.

[3] Singh, K. (2005). Final Annual Report Development of sustainable Farming system model for the Irrigated Agro-Ecosystem of Eastern UP, ICAR, Ad-hoc project, Department of Agronomy, I.Ag.Sci., B.H.U., Varanasi. India

[4] Subbiah, B.V. and Asija, G.L. (1973). A rapid procedure for estimation of available nitrogen in soils. *Current Sciences*, 28(8): 259-260.

[5] Bhargava, B.S. and Raghuipathi, H.B. (1993). Analysis of plant materials for macro and micronutrients "Methods of Analysis of Soils, Plants, Waters and Fertilizers"

edited by Tandon, H.L.S., Fertilizer Development and Consultation Organization, pp. 49-82.

[6] Sriram, C.; Thyagraj, C.R.; Mayande, V.M.; and Srinivas Rao, P. (1999). Indo-U.S. Project on Research of Dryland Agriculture (Operation Search). Central Research Instt. for Dryland Agriculture, Hyderabad, India. *Annexure I.*

[7] Singh, B.; Singh, Y.; and Meelu, O.P. (1996). Management of soil organic matter for maintained productivity in rice-wheat in India. In: Studies in Indian Agro-ecosystems (Pathak, P.S. and Gopal, B. eds.), National Institute of Ecology, p. 85-90.

[8] Yadav R.L.; Dwivedi, B.S.; and Pandey, P.S. (2000). Rice-wheat cropping system: Assessment of sustainability under green manuring and chemical fertilizer inputs. *Field Crops Res.*, 65: 15-30.

[9] Sharma, R.P.; Pathak, S.K.; Haque, M.; and Raman, R.R. (2004). Diversification of traditional rice (*Oryza sativa*) based cropping system for sustainable production in south Bihar alluvial plains. *Indian J. Agron.*, 49(4): 218-222.

[10] Yadav, M.P.; Rai, J.; Kushwaha, S.P.; and Singh, G.K. (2005). Production potential and economic analysis of various cropping systems for central plains zone of Uttar Pradesh. *Indian J. Agron.*, 50(2): 83-85.

[11] Saroch, K.; Bhargava, M.; and Sharma, J.J. (2005). Diversification of existing rice (*Oryza sativa*) based cropping system for sustainable productivity under irrigated conditions. *Indian J. Agron.*, 50(2): 86-88.

[12] Newaj, Ram and Yadav, D.S. (1992) Production potential and labour employment under different cropping system under upland conditions of eastern Uttar Pradesh. *Indian J. Agron.*, 37(3): 401-406.

[13] Urkurkar, J.S; Chitale, S.; and Tiwari, F. (2010). Effect of organic v/s chemical nutrient packages on productivity, economics and physical status of soil in rice–potato cropping system in Chhattisgarh. *Indian J. Agron.*, 55 (1): 6-10

[14] Hati, K.M.; Mandal, K.G.; Mishra, A.K.; Ghosh, P.K.; and Acharya, C.L. (2001). Effect of irrigation regimes and nutrient management on soil water dynamics, evapo-transpiration and yield of wheat in vertisols. *Indian Journal of Agricultural Sciences*, 71(9): 581-587.

[15] Wu, S.; Ingham, E.; and Hu, D. (2002). In: 17th World Congress of Soil Science, Queen Sirkit National Convention Centre, 14-21 August 2002, Bangkok, Thailand, Abstract Vol. V, p. 1756.

2

Organic Weed Control and Cover Crop Residue Integration Impacts on Weed Control, Quality, Yield and Economics in Conservation Tillage Tomato

Andrew J. Price, Leah M. Duzy, Kip S. Balkcom, Jessica A. Kelton, Ted S. Kornecki and Lina Sarunaite

Additional information is available at the end of the chapter

Abstract

The increased adoption of conservation tillage and organic weed control practices in vegetable production requires more information on the role of various cover crops in integrated weed control, tomato quality, and yield. Two conservation-tillage systems utilizing crimson clover and cereal rye as winter cover crops were compared to a conventional black polythene mulch system, with or without organic weed management options, for weed control, tomato yield, and profitability. All cover crops were terminated with a mechanical roller/crimper prior to planting. Organic weed control treatments included: 1) flaming utilizing a one burner hand torch, 2) PRE application of corn gluten, 3) PRE application of corn gluten followed by flaming, or 4) intermittent hand weeding as needed. A non-treated control and a standard herbicide program were included for comparison. The herbicide program consisting of a PRE application of S-metolachlor (1.87 kg a.i./ha) followed by an early POST metribuzin (0.56 kg a.i. /ha) application followed by a late POST application of clethodim (0.28 kg a.i./ha). In general, high-residue clover and cereal rye cover crops provided substantial suppression of Palmer amaranth, large crabgrass, and yellow nutsedge. Across systems, minimum input in high-residue systems provided the highest net returns above variable costs compared to organic herbicide treatments that are costly and provide marginal benefit.

Keywords: Conservation agriculture, cover crop, fruit

1. Introduction

In recent years, growing concerns over the environmental impact of conventional agricultural practices, coupled with a surge in consumer demand for sustainably-produced products, have led to increased grower adoption of organic agriculture. In 2011, cropland in the United States (U.S.) dedicated to organic vegetable production totaled over 47 thousand ha [1]. Organically produced vegetable sales, were estimated at 1.07 billion USD in 2011 [1]. Given the steady rise in organic product interest and efforts to ensure agricultural sustainability, a substantial amount of research has been dedicated to organic fruit and vegetable production in order to guarantee successful adoption of these practices as an alternative to conventional agriculture.

Unlike conventional agricultural practices, an organic approach to agriculture eliminates the use of synthetic pesticides and fertilizers and, instead, relies on biological and cultural pesticide control and organic soil amendments such as manure and crop residue to maintain soil fertility [2]. The goal of organic agriculture includes producing food and fiber products in a manner that increases biodiversity, promoting soil health, and reducing environmental degradation due to agricultural practices. A number of ecological differences have been noted in previous research when comparing conventional and organic agriculture [3,4]. Comparisons of soil properties and pest population dynamics for organic and traditional farming practices note differences between these systems that affect the agroecosystem [3,4].

2. Case study

In the U.S. approximately 1.36 million tons of in the open, fresh market tomatoes, worth over 1.134 billion USD, were produced on nearly 41.2 thousand ha in 2014 [5]. Tomato production systems typically utilize conventional tillage, a bedded plastic mulch culture, and multiple herbicide applications to control weeds. These conventional tillage systems enhance soil erosion and nutrient loss by reducing rainfall infiltration [6]. Additionally, tillage increases aeration which increases the rate of organic matter mineralization in the surface soil, thus reducing soil organic matter content, soil cation exchange capacity and potential productivity [7, 8].

Plastic mulch can increase soil temperature which can expedite tomato harvest [9]. Tomato harvest was not early following a hairy vetch mulch system [10, 11]. The use of plastic mulches in sustainable or organic production systems is in question by some producers and consumers since the mulch itself is non-biodegradable and made of non-renewable resources. Another environmental disadvantage with using plastic mulch vs. organic mulches is increased chemical runoff from plastic mulch systems and subsequent offsite chemical loading [12]. Thus, the intensive use of pesticides in vegetable production has resulted in ecological concerns. Therefore, alternative production practices that reduce tomato production inputs while maintaining yield and quality are desired.

One alternative for alleviating the aforementioned concerns is the use of high residue cover crops combined with reduced tillage. Cover crops in conservation-tillage systems can be terminated during early reproductive growth by mechanically rolling and treating with burndown herbicides to leave a dense mat of residue (> 4,500 kg/ha) on the soil surface into which cash crops are planted [13, 14]. Adoption of high residue cover crops is increasing in southeastern U.S. corn (*Zea mays* L.) and cotton (*Gossypium hirsutum* L.) row crop systems [15, 16, 17, 18, 19, 20]. Because the southeastern U.S. typically receives adequate rainfall in the winter months, timely planted winter cover crops can attain relatively high maturity and biomass before termination. Cover crops can enhance the overall productivity and soil quality by increasing organic matter and nitrogen content [21], as well as aid in water conservation by increasing soil water infiltration rates [22]. Additionally, previous research has also focused on weed control provided by high residue cover crops in both field and vegetable crops [23, 24, 25].

Winter cover crop biomass can affect subsequent early season weed control [26, 27]. Cover crop residue facilitates weed control by providing an unfavorable environment for weed germination and establishment under the residue as well as allelopathy [28, 29]. Teasdale and Daughtry [30] reported 52–70% reduction in weed biomass with live hairy vetch cover crop compared to a fallow treatment owing to changes in light and soil temperature regimen under the vetch canopy. Teasdale and Mohler [27] reported that legume mulches such as crimson clover and hairy vetch (*Vicia villosa* Roth) suppressed redroot pigweed (*Amaranthus retrofloxus* L.) at an exponential rate as a function of residue biomass.

However, adoption of cover crops in tomato production has been limited because (1) currently available transplanters have problems penetrating heavy residue and (2) heavy cover crop residue can intercept delivery of soil-active herbicides. Research in the last two decades has extensively debated the advantages and disadvantages of cover crops vs. conventional plastic mulch systems for tomato production. Better or comparable tomato yields with hairy vetch cover crop system have been reported compared to the conventional polyethylene mulch system [31, 32]. Akemo et al. [33] also reported higher tomato yield with spring sown cover crops than the conventionally cultivated check. However, weed control with cover crops varies with cover crop species, amount of residue produced, and environmental conditions. Teasdale [28] reported that biomass levels achieved by cover crops before termination was sufficient only for early season weed control. Supplemental weed control measures are usually required to achieve season long weed control and to avoid yield losses [34, 23].

Cereal rye and crimson clover are two common winter cover crops widely used in the southeastern U.S. Both cover crops contain allelopathic compounds and produce residues that inhibit weed growth [15, 29, 35]. Brassica cover crops are relatively new in the southeastern U.S. but are becoming increasingly popular due to their potential allelopathic effects. Therefore, the objectives of this research were to evaluate: 1) weed control in two different high residue cover crop conservation tillage systems utilizing the Brazilian [13] high residue cover crop management system including cover crop rolling and 2) tomato stand establishment, yield, and net returns of conservation-transplanted tomatoes compared to the polythene mulch system following three different organic herbicide management systems.

3. Materials and methods

Field Experiment. The experiment was established in autumn 2006 at the North Alabama Horticulture Experiment Station, Cullman, AL on a Hartsells fine sandy loam soil (Fine-loamy, siliceous, sub-active, thermic Typic Hapludults). The experimental design was a randomized complete block with four replicates. Plot size at both locations was 1.8 by 6 m containing a single row of tomatoes with a 0.5 m spacing between plants.

The two winter cover crops (cereal rye cv Elbon and crimson clover cv AU Robin) were compared to black polythene mulch for their weed suppressive potential and effect on yield and grade of fresh market tomatoes. Winter cover crops were planted with a no till drill in the fall. Rye was seeded at a rate of 100 kg/ha, whereas clover was seeded at 28 kg/ha. Since the overall objective was to evaluate weed control practices, general production practices included staking, traditional plant pest and plant pathogen methods, and fertilization was utilized to exclude any other pest and fertilization interactions and is a limitation of this case study. Nitrogen was applied at a rate of 67 kg/ha on rye plots in early spring of each year. Cover crops were terminated at flowering stage in late spring. To determine winter cover crop biomass production, plants were clipped at ground level from one randomly selected 0.25 m^2 area per replicate immediately before termination. Plant samples were dried at 65 C for 72 hours and weighed. Cover crops were terminated with a mechanical roller crimper prior to an application of glyphosate at 1.12 kg a.e. /ha^{-1}. The rolling process produced a uniform residue cover over the plots.

All three systems (two winter cover crops plus plastic mulch) were evaluated with and without herbicide for weed control. Organic weed control treatments included: 1) flaming utilizing a one burner hand torch, 2) PRE application of corn gluten, 3) PRE application of corn gluten followed by flaming, or 4) intermittent hand weeding as needed. A non-treated control and a standard herbicide program were included for comparison. The herbicide program consisting of a PRE application of S-metolachlor (1.87 kg a.i. ha^{-1}) followed by an early POST (EPOST) metribuzin (0.56 kg a.i. ha^{-1}) application followed by a late POST (LPOST) application of clethodim (0.28 kg a.i.ha^{-1}). The PRE corn gluten application occurred immediately after tomato transplanting while the PRE herbicide application occurred prior to placing the plastic on top of the beds, the EPOST application was applied two weeks after transplanting, and the LPOST application was delayed until tomatoes were near mid-bloom. Flaming and hand hoeing was accomplished one week after transplanting and subsequently every two weeks following until harvest. Tomato cv. 'Florida 47' seedlings were transplanted on April 12, 2007.

Tomato seedlings were planted with a modified RJ no-till transplanter (RJ Equipment, Blenhiem, Ontario, Canada), which included a subsoiler shank installed to penetrate the heavy residue and disrupt a naturally occurring compacted soil layer found at both experimental sites at a depth of 30-40 cm. Additionally, two driving wheels were utilized (one wheel on each side of the tomato row) instead of the original single wheel at the center of the row, to improve stability and eliminate drive wheel re-compaction of the soil opening created by the shank. The plastic-mulch plots were conventionally tilled utilizing a tractor mounted rototiller prior to bedding and plastic installation; tomatoes were hand transplanted in the plastic mulch each year. Water was applied to all the plots immediately after transplanting. Thereafter, plots were irrigated every other day using a surface drip tape. Fertilizer 13-13-13 was applied prior to planting achieving 448 kg of N/ha^{-1} and then 7.8 kg of calcium nitrate ha^{-1} was applied once every week with the irrigation system.

Weed control was evaluated by visual ratings (0% = no control, 100% = complete control) 28 days after treatment (DAT) of the EPOST herbicide application. All weed species present were evaluated for control (as a reduction in total above ground biomass resulting from both reduced emergence and growth). Stand establishment was determined by counting the number of living tomato plants in each plot two weeks after LPOST application. Ripe tomatoes were hand harvested from the entire plot area in weekly intervals and sorted according to size (small, medium, large, and extra large categories).

Statistical Analysis. Non-normality and heterogeneous variances were encountered with percent control data. Various approaches were tried to alleviate these statistical problems and the arcsine transformation was deemed the best compromise between achieving normality of residuals and among treatment homogeneity of variances. The transformed data were subjected to mixed models analysis of variance as implemented in JMP statistical software. Years, organic herbicide treatments and ground cover treatments were considered fixed effects while their interaction with treatment replication was considered random effects. Differences between treatments means were determined by Fisher's protected LSD ($\alpha = 0.05$).

Economic analysis. Net returns above variable treatment costs (NRAVTC) were estimated as the difference between revenues and variable treatment costs (US$ ha^{-1}). The average weekly dollar per box (assuming an 11.34 kg box^{-1}) price for the four harvest weeks was used to calculate revenue by grade (i.e., small, medium, large, and extra-large). The weekly prices were from domestic suppliers at the terminal market in Atlanta, Georgia [36]. Low- and high-end prices from 2007 were reported for each grade category from suppliers (domestic suppliers aggregated by State), excluding international suppliers. The low-end and high-end tomato prices by size were the average of prices in 2007 across suppliers, and are presented in Table 1. All prices were reported in 2007 US$.

Tomato Size	Low-end Price	High-end Price	Mean
	US$ box^{-1}		
Small	10.06	10.69	10.38
Medium	9.47	10.14	9.81
Large	9.34	9.99	9.67
Extra-large	9.41	10.28	9.85
Mean	9.57	10.58	

Table 1. Tomato prices by size by low-end and high-end price.

The average marketing year price, regardless of organic certification, received by producers in Alabama in 2007 for fresh market tomatoes across all sizes (7.21 US$ box^{-1}). For organically produced tomatoes, the average price received by Alabama producers for organic tomatoes in 2008 of 9.32 US$ box^{-1} across all sizes [37]. Data for organic tomatoes was not available in 2007. Therefore, the low-end prices by size were used in the analysis.

Productions costs for the three covers and five weed control treatments were adapted from 2008 tomato enterprise budgets [38] and experiment specific treatment costs. A partial budgeting approach was used to calculated variable treatments costs; therefore, the only costs considered were costs that differed by treatment and costs that varied by yield (Table 2). Costs that vary by yield include harvest costs, as well as grading and packing labor costs. Fixed costs, such as management costs, rent, and depreciation on machinery and buildings, differ by operation; therefore, they were not included in the analysis.

Weed Control	Cover Type		
	Plastic	Rye	Clover
	US$ ha^{-1}		
No Treatment	2226	505	376
Handweed	3658	1937	1808
Flame Corn Gluten	12935	11214	11085
Flame	2859	1138	1009
Herbicide	2392	671	542

Table 2. Variable treatment costs (excluding costs that vary by yield).

4. Results and discussion

Cover Crop Biomass. The quantity of cover crop biomass produced at both locations differed among cover crops, with rye producing 9363 kg/ha, and crimson clover producing 5481 kg/ha of dry matter.

Weed Control. The major weeds in the cover crop and plastic mulch plots included Palmer amaranth (*Amaranthus Palmeri* L.), large crabgrass (*Digitaria sanguinalis* L.), and yellow nutsedge (*Cyperus esculentus* L.).

Palmer amaranth. Early *Palmer amaranth* control averaged over weed management systems, clover and rye cover treatments provided excellent *Palmer amaranth* control (90 and 96% respectively) compared to the conventional plastic system (5% control) (Table 3). The plastic system provides some inherent weed control regardless of additional inputs, however, it provided no weed control in the punched holes and the area adjacent the bed. *Palmer amaranth* control in clover utilizing corn gluten and flaming was equivalent to the clover plus herbicide standard. *Palmer amaranth* in rye utilizing all organic methods excluding hand weeding provided weed control equivalent to the rye plus herbicide standard. Late *Palmer amaranth* control ratings generally remained stable except increases for plastic due to the inherent control discussed above.

Large Crabgrass. Early crabgrass control averaged over weed management system reflected control similar to Palmer amaranth, clover and rye cover treatments provided excellent crabgrass control (92 and 98% respectively) compared to the conventional plastic system (5% control) (Table 4). All rye systems provided excellent control. Late season crabgrass control was generally higher than that of *Palmer amaranth*.

Yellow nutsedge. Early yellow nutsedge control averaged over weed management systems reflected control similar to Palmer amaranth and large crabgrass with clover systems providing an average 93% control and rye systems providing an average 95% control. Control in both clover and rye systems was excellent regardless of treatment revealing that winter cover crops suppress nutsedge in high-residue systems.

| | % Weed Control | | | | | |
| | Early Control | | | Late Control | | |
Cover	Pigweed	Crabgrass	Nutsedge	Pigweed	Crabgrass	Nutsedge
Clover	90[a]	92[a]	93[a]	92[a]	98[a]	98[a]
Rye	96[a]	98[a]	95[a]	88[a]	97[a]	98[a]
Plastic	5[b]	5[b]	5[b]	33[b]	37[b]	43[b]
LSD (α = 0.10)	7	13	9	12	14	13
Weed Control[1]						
1	63[ba]	64[a]	63[ba]	60[b]	71[a]	73[ba]
2	57[b]	61[a]	64[ba]	73[ba]	81[a]	82[ba]
3	61[ba]	61[a]	55[b]	77[ba]	80[a]	82[ba]
4	65[ba]	65[a]	66[ba]	61[b]	65[a]	66[b]
5	72[a]	72[a]	74[a]	86[a]	87[a]	96[a]
LSD (α = 0.10)	10	10	12	15	18	17
Combination						
Clover 1	93[a]	96[a]	90[a]	88[a]	97[a]	98[a]
Clover 2	80[a]	86[a]	93[a]	92[a]	98[a]	98[a]
Clover 3	85[a]	85[a]	86[a]	91[a]	98[a]	99[a]
Clover 4	97[a]	97[a]	99[a]	92[a]	97[a]	98[a]
Clover 5	97[a]	96[a]	97[a]	99[a]	99[a]	98[a]
Plastic 1	0[b]	0[b]	0[b]	6[b]	20[bc]	23[b]

| | % Weed Control | | | | | |
| | Early Control | | | Late Control | | |
Cover	Pigweed	Crabgrass	Nutsedge	Pigweed	Crabgrass	Nutsedge
Plastic 2	0[b]	0[b]	0[b]	49[ba]	50[bac]	50[ba]
Plastic 3	0[b]	0[b]	0[b]	50[ba]	50[bac]	50[ba]
Plastic 4	0[b]	0[b]	0[b]	0[b]	0[c]	0[b]
Plastic 5	23[b]	23[b]	25[b]	61[a]	65[ba]	90[a]
Rye 1	97[a]	97[a]	98[a]	86[a]	97[a]	99[a]
Rye 2	92[a]	97[a]	98[a]	79[a]	96[a]	99[a]
Rye 3	97[a]	99[a]	81[a]	90[a]	94[a]	96[a]
Rye 4	98[a]	99[a]	99[a]	90[a]	98[a]	99[a]
Rye 5	96[a]	98[a]	99[a]	98[a]	99[a]	99[a]
LSD (α = 0.10)	17	17	21	27	31	29

[1]Weed control methods are as follows: (1) non-treated; (2) hand-weeded; (3) corn gluten + flame; (4) flame; and (5) herbicide.

Table 3. Weed Response to Cover Crops and Weed Control Methods – North Alabama Horticultural Research Center 2007.

Yield

Aside from the herbicide treatment, greater than 20% of the total tomato yield were cull tomatoes under plastic cover.

| | Tomato Yield (kg/ha) | | | | | | |
Cover	Cull	S	M	L	XL	Total	Market[2]
Clover	5577[a]	4838[a]	9906[a]	12298[a]	263[a]	32883[a]	27305[a]
Rye	5479[a]	4778[a]	9649[a]	11031[a]	272[a]	31210[a]	25731[a]
Plastic	4226[b]	2599[b]	4566[b]	7526[b]	158[a]	19074[b]	14848[b]
LSD (α = 0.10)	612	576	1078	1931	197	3254	2931
Weed Control[1]							
1	4159[c]	4006[a]	6669[b]	7149[c]	283[ba]	22266[c]	18107[c]
2	5112[bac]	4634[a]	8220[b]	8466[cb]	54[b]	26486[cb]	21374[cb]
3	5554[ba]	4003[a]	8355[b]	11248[b]	241[ba]	29402[b]	23848[b]
4	4547[bc]	3871[a]	6471[b]	6565[c]	58[b]	21512[c]	16966[c]
5	6098[a]	3845[a]	10486[a]	17996[a]	518[a]	38944[a]	32846[a]

	Tomato Yield (kg/ha)						
Cover	Cull	S	M	L	XL	Total	Market[2]
LSD (α = 0.10)	790	744	1392	2493	255	4201	3784
Combination							
Clover 1	5076[bac]	4972[bdac]	9197[bdac]	10390[bedc]	240[a]	29874[bc]	24799[bc]
Clover 2	6204[a]	6395[a]	10218[bdac]	10004[bedc]	161[a]	32982[bac]	26779[bac]
Clover 3	5673[ba]	5315[bac]	10814[bac]	11284[bc]	194[a]	33280[bac]	27608[bac]
Clover 4	4233[bac]	381[ebdc]	7463[bdc]	8029[edc]	125[a]	23660[edc]	19427[dc]
Clover 5	6702[a]	3698[ebdc]	11838[ba]	21782[a]	594[a]	44615[a]	37913[a]
Plastic 1	2974[c]	2107[e]	2226[e]	2629[ed]	0[a]	9937[e]	6963[d]
Plastic 2	4556[bac]	2676[ed]	5953[de]	8388[edc]	0[a]	21574[edc]	17018[dc]
Plastic 3	5098[bac]	2838[edc]	5693[de]	10491[bdc]	238[a]	24357[dc]	19259[dc]
Plastic 4	3494[bc]	2143[e]	2668[e]	1892[e]	0[a]	10197[ed]	6703[d]
Plastic 5	5006[bac]	3229[ebdc]	6289[dec]	14228[bac]	552[a]	29304[bc]	24297[bc]
Rye 1	4428[bac]	4937[bdac]	8584[bdc]	8429[edc]	610[a]	26988[c]	22560[c]
Rye 2	4577[bac]	4831[bdac]	8490[bdc]	7005[edc]	0[a]	24902[c]	20325[c]
Rye 3	5892[ba]	3855[ebdc]	8559[bdc]	11970[bc]	292[a]	30567[bac]	24676[bc]
Rye 4	5913[ba]	5659[ba]	9283[bdac]	9775[bedc]	50[a]	30679[bac]	24767[bc]
Rye 5	6587[a]	4608[ebdac]	13332[a]	17978[ba]	409[a]	42913[ba]	36327[ba]
LSD (α = 0.10)	1368	1288	2410	4319	441	7277	6554

[1]Weed control methods are as follows: (1) non-treated; (2) hand-weeded; (3) corn gluten + flame; (4) flame; and (5) herbicide.

[2]Market is the marketable yield calculated by subtracting the culls from the total.

Table 4. Tomato Yield Response to Cover Crops and Weed Control Methods - North Alabama Horticultural Research Center 2007.

Economics

All treatments produced numerically higher NRVTC than the control, with the exception of plastic cover with flame treatment (Table 5). The clover cover and herbicide treatment produced the highest NRAVTC in 2007, followed by rye cover and herbicide treatment (Table 6). Both the non-treated control combined with clover and rye, as well as flame and hand-weeded treatments with clover cover, yielded higher NRAVTC than plastic with herbicide treatment. Across all cover treatments, corn gluten + flame had the lowest NRAVTC. The performance of corn gluten + flame was directly related to the cost of the corn gluten. As discussed above the corn gluten + flame weed control with clover cover had the third highest market tomato yields.

While total market yield is an important indicator of net returns, the distribution of tomatoes by size determines the level of revenue depending on the price by size. The price for each size is driven by the supply of each type of size and when the tomatoes are harvested during the season. This analysis did not consider harvest period in the revenue determination.

Cover Type	Weed Control[1]	NRAVTC[2]		Difference from Control[3]
		Mean	SD	
		(US$ ha⁻¹)		
Clover	1	4680	1568	2254
	2	3718	1524	1293
	3	-5465	702	-7890
	4	2951	1526	525
	5	6910	1167	4485
Plastic	1	-769	421	-3194
	2	-245	2079	-2671
	3	-9088	1809	-11513
	4	-1439	480	-3865
	5	2426	549	0
Rye	1	4130	625	1704
	2	2262	651	-164
	3	-6261	1024	-8686
	4	3954	1663	1528
	5	6563	261	4137

[1] Weed control methods are as follows: (1) non-treated; (2) hand-weeded; (3) corn gluten + flame; (4) flame; and (5) herbicide.

[2] Net returns above variable treatment cost (NRAVTC); standard deviations are shown in parentheses.

[3] The control is plastic cover with no weed control.

Table 5. Net returns above variable treatment costs by treatment and the difference between treatments and the control.

This research demonstrates that high residue cover crops like cereal rye and clover can provide improved weed control compared to black polyethylene mulch. Previous research has also reported improved weed control with increased biomass production by cover crops [39]. Increased weed control has also been observed by Nagabhushna et al. [40] with an increase in the seeding rate of rye. Another important factor which could have facilitated increased weed control by rye and clover residue is rolling with mechanical roller crimper. The rolling process resulted in a uniform mat of residue on the soil surface that was a substantial physical barrier for weed seedlings to emerge through compared to tomato plant openings in the plastic mulch system that provides no barrier. Yenish et al. [41] also reported inconsistent control with cover crop residue and concluded herbicides were always required to achieve optimum weed control in corn. However, Yenish et al. cautioned weed control should not be the only criterion in selection of cover crops. Factors like cost and ease of establishment, impact on yield should be taken into consideration before selecting a cover crop. Results in this paper are short term effects of converting from a conventional plastic mulch system to two high-residue conservation tillage systems. These results indicate the economic possibility of growing fresh market tomatoes utilizing a conservation tillage system while maintaining yields and economic returns. However, the long term impact of these systems on yield and profitability require further investigation.

Author details

Andrew J. Price[1*], Leah M. Duzy[1], Kip S. Balkcom[1], Jessica A. Kelton[2], Ted S. Kornecki[1] and Lina Sarunaite[3]

*Address all correspondence to: andrew.price@ars.usda.gov

1 United States Department of Agriculture, Agricultural Research Service, National Soil Dynamics Laboratory, Auburn, Alabama, USA

2 Auburn University, Auburn, Alabama, USA

3 Institute of Agriculture, Lithuanian Research Centre for Agriculture and Forestry, Lithuania

References

[1] United States Department of Agriculture. 2015. Quick States 2.0. http://quick-stats.nass.usda.gov/.

[2] Lammerts van Bueren, E. T., S. S. Jones, L. Tamm, K. M. Murphy, J. R. Myers, C. Leifert, and M. M. Messmer. 2011. The need to breed crop varieties suitable for organic farming, using wheat, tomato and broccoli as examples: A review. NJAS-Wageningen Journal of Life Sciences. 58:193-205.

[3] Madden, N. M., J. P. Mitchell, W. T. Lanini, M. D. Cahn, E. V. Herrero, S. Park, S. R. Temple, and M. van Horn. 2004. Evaluation of conservation tillage and cover crop systems for organic processing tomato production. HortTechnology 14:243:250.

[4] Drinkwater, L. W., D. K. Letourneau, F. Workneh, A. H. C. van Bruggen, and C. Shennan. 1995. Fundamental differences between conventional and organic tomato agroecosystems in California. Ecological Applications. 5:1098-1112.

[5] United States Department of Agriculture. 2015. Quick States 2.0. http://quick-stats.nass.usda.gov/. Accessed: July 28, 2015.

[6] Blough, R. F., A. R. Jarrett, J. M. Hamlett and M. D. Shaw. 1990. Runoff and erosion rater from silt, conventional, and chisel tillage under simulated rainfall. Transactions of ASAE. 33:1557–1562.

[7] Franzluebbers, A.J., G.W. Langdale, and H.H.Schomberg. 1999. Soil carbon, nitrogen, and aggregation in response to type and frequency of tillage. Soil Sci. Soc. Am. J. 63:349–355.

[8] Mahboubi, A.A., R. Lal, and N.R. Faussey. 1993. Twenty-eight years of tillage effects on two soils in Ohio. Soil Sci. Soc. Am. J. 57:506–512.

[9] Teasdale, J.R.and A.A. Abdul-Baki. 1995. Soil temperature and tomato growth associated with black polythene and hairy vetch mulches. J. Amer. Soc. Hort. Sci. 120:848-853.

[10] Abdul-Baki. A.A., J.R. Teasdale, R. Korcak, D.J. Chitwood, and R.N. Huettel. 1996. Fresh-market tomato production in a low-input alternative system using cover crop mulch. HortScience. 31:65-69.

[11] Teasdale, J.R.and A.A. AbdulBaki. 1997. Growth analysis of tomatoes in black plastic and hairy vetch production systems. Hortscience. 32:659-663.

[12] Arnold, G. L., M. W. Luckenbach, and M. A. Unger. 2004. Runoff from tomato cultivation in the estuarine environment: biological effects of farm management practices. J Exp Marine Biol and Ecol. 2:323-346.

[13] Derpsch, R., C. H. Roth, N. Sidiras, and U. Köpke. 1991. Controle da erosão no Paraná, Brazil: Sistemas de cobertura do solo, plantio directo e prepare conservacionista do solo. Deutsche Gesellschaft für Technische Zusammenarbeit (GTZ) GmbH, Eschborn, SP 245, Germany.

[14] Reeves, D.W. 2003. A Brazilian model for no-tillage cotton production adapted to the southeastern USA. Proc. II World Congress on Conservation Agriculture- Producing in Harmony with Nature. Iguassu Falls, Paraná, Brazil. Aug 11-15, 2003:372-374.

[15] Price, A.J., C. D. Monks, A. S. Culpepper, L. M. Duzy, J. A. Kelton, M. W. Marshall, L. E. Steckel, L.M. Sosnoskie and R. L. Nichols. High Residue Cover Crops Alone or with Strategic Tillage to Manage Glyphosate-Resistant Palmer amaranth (Amaranthus palmeri) in Southeastern Cotton (Gossypium hirsutum). Journal of Soil and Water Quality. (in press).

[16] Aulakh, J.S. M., Saini, A.J. Price, W.H. Faircloth, E. van Santen, G.R. Wehtje, and J.A. Kelton. 2015. Herbicide and Rye Cover Crop Residue Integration Affect Weed Control and Yield in Strip-Tillage Peanut. Peanut Sci. 42:30-38.

[17] Price, A. J., K. S. Balkcom, L. M. Duzy, and J. A. Kelton. 2012. Herbicide and cover cop residue integration for Amaranth control in conservation agriculture cotton. Weed Technol. 26:490-498

[18] Price, A.J., D.W. Reeves, and M.G. Patterson. 2006. Evaluation of weed control provided by three winter cereals in conservation-tillage soybean. Renewable Agric. and Food Systems. 21:159-164.

[19] Reeves, D.W., A.J. Price, and M.G. Patterson. 2005. Evaluation of three winter cereals for weed control in conservation-tillage nontransgenic cotton. Weed Technol. 19: 731-736.

[20] Sainju, U.M., and B.P. Singh. 2001. Tillage, cover crop, and kill-planting date effects on corn yield and soil nitrogen. Agron. J. 93: 878–886

[21] Sainju, U.M., B.P. Singh, and W.F. Whitehead. 2002. Long-term effects of tillage, cover crops, and nitrogen fertilization on organic carbon and nitrogen concentrations in sandy loam soils in Georgia, USA. Soil Till. Res. 63:167-179.

[22] Arriaga, F.J. and K.S. Balkcom. 2006. Benefits of conservation tillage on rainfall and water management. In: Hatcher, K. J., editor. Proceedings of the 2005 Georgia Water Resources Conference, April 25-27, 2005.

[23] Teasdale, J.R.and A.A. Abdul-Baki. 1998. Comparison of mixtures vs. monocultures of cover crops for fresh-market tomato production with and without herbicide. Hortscience. 33:1163-1166.

[24] Creamer, N.G., M.A. Bennett, and B.R. Stinner. 1997. Evaluation of cover crop mixtures for use in vegetable production systems. HortScience. 32:866-870.

[25] Price, A. J. and J. K. Norsworthy. 2013. Cover crop use for weed management in Southern reduced-tillage vegetable cropping systems. Weed Technol. 27:212-217.

[26] Saini, M., A. J. Price, and E. van Santen. 2006. Cover crop residue effects on early-season weed establishment in a conservation-tillage corn-cotton rotation. 28th Southern Conservation Tillage Conference 28:175-178.

[27] Teasdale, J.R.and C.L. Mohler. 2000. The quantitative relationship between weed emergence and the physical properties of mulches. Weed Sci. 48:385-392.

[28] Price A.J., M.E. Stoll, J.S. Bergtold, F.J. Arriaga, K.S. Balkcom, T.S. Kornecki, and R.L. Raper. 2008. Effect of cover crop extracts on cotton and radish radicle elongation. Comm. Biometry Crop Sci. 3:60-66.

[29] Teasdale, J.R. 1996. Contribution of cover crops to weed management in sustainable agricultural systems. J. Prod. Agric. 9:475-479.

[30] Teasdale JR & Daughtry CST (1993) Weed control by live and desiccated hairy vetch (Vicia villosa). Weed Science 41, 207 212

[31] Abdul-Baki A.A., and J.R. Teasdale. 1993. A no-tillage tomato Production system using hairy vetch and subterranean clover mulches. HortScience. 28:106-108.

[32] Abdul-Baki, A.A., J.R. Teasdale, R.W. Goth, and K.G. Haynes. 2002. Marketable yields of fresh-market tomatoes grown in plastic and hairy vetch mulches. HortScience. 37:878-881.

[33] Akemo, M.C., M.A. Bennett, and E.E. Regnier. 2000. Tomato growth in spring-sown cover crops. HortScience. 35:843-848.

[34] Masiunas, J.B., L.A. Weston, and S.C. Weller. 1995. The impact of rye cover crops on weed populations in a tomato cropping system. Weed Sci. 43:318-323.

[35] Barnes, J.P. and A.R. Putnam. 1983. Rye residues contribute weed control in no-tillage cropping systems. J. of Chem. Ecol. 9:1045-1057.

[36] USDA. 2015. Fruit and Vegetable Market News. Agricultural Marketing Service, United States Department of Agriculture (USDA). Available at Web site https:// www.marketnews.usda.gov/mnp/fv-home (verified August 3, 2015).

[37] USDA. 2015. Quick Stats. National Agricultural Statistics Service, United States Department of Agriculture (USDA). Available at Web site http://quick-stats.nass.usda.gov/ (verified August 3, 2015).

[38] MSU. 2007. Traditional and organic vegetables 2008 planning budgets. Budget Report 2007–08. Department of Agricultural Economics, Mississippi State University (MSU). Available at website http://www.agecon.msstate.edu/whatwedo/budgets/archive.asp (verified August 3, 2015)

[39] Mohler, C. L. and J. R. Teasdale. 1993. Response of weed emergence to rate of Vicia villosa Roth and Secale cereale L. residue. Weed Res. 33:487–499.

[40] Nagabhushana, G.G., A.D. Worsham, and J.P. Yenish. 2001. Allelopathic cover crops to reduce herbicide use in sustainable agricultural systems. Allelopathy J. 8:133-146.

[41] Yenish, J.P., A.D. Worsham, and A.C. York. 1996. Cover crops for herbicide replacement in no-tillage corn (Zea mays). Weed Technol. 10:815-821.

3

Organic Dairy Sheep Production Management

Juan C. Angeles Hernandez, Octavio A. Castelan Ortega,
Sergio Radic Schilling, Sergio Angeles Campos, A. Hilda Ramirez Perez and
Manuel Gonzalez Ronquillo

Additional information is available at the end of the chapter

Abstract

Organic production systems are based on natural processes, the use of local feed resources, and the maintenance of biodiversity in all senses. Several studies have noted the positive effects of organic sheep milk production systems on animal welfare, animal health, product quality, and environmental impact. On the other hand, it has been reported that dairy sheep organic farms show lower milk yields and increase the susceptibility to environmental impacts compared with conventional farms. The standards that regulate feeding management in organic systems are one of the most critical factors that influence milk production performance. Lower milk production is also associated with poor ability to adapt specialized dairy breeds to organic management, low genetic potential for milk production in native and local breeds, and elevated dependence on environmental conditions. However, the aim of organic dairy production is not to reach maximum dairy productivity but rather to integrate animal and crop production and to develop a symbiotic relationship between recyclable and renewable resources; furthermore, organic production positively affects the employment rate and quality of life in rural communities. Organic dairy sheep production is one means of improving the balance between society's demand for food and the ecological impact of the agro-alimentary industry.

Keywords: Sheep, milk production, organic system, sustainability

1. Introduction: A brief overview of organic farming

Society's demand for foodstuffs is growing at a higher rate than current levels of production due to population growth and the rise in average income. According to the FAO, "food security exists when all the people, at all times, have physical, social and economic access to sufficient, safe and nutritious food." Over the last few years, some consumers have expressed increasing concern regarding the origins of their food, its social and ecological impacts, and the fairness of its production. These customers prefer organic products, based on their perception that organic farming generates benefits associated with animal welfare, food quality, food safety, environmental concerns, and community development [1].

Due to its agro-ecological and holistic approaches and the competitive prices for organic products in the market, organic farming has developed into a small but important sector in agricultural production [2]. In 2012 alone, the "organic market" was worth approximately 50 billion euros. The International Federation of Organic Agriculture Movements (INFOAM) [3] reported that in 2012 some 37.5 million hectares of land were dedicated to organic agriculture, which represented 0.87% of total agricultural land. Australia is the country with the largest area used in organic agriculture, with 12 million hectares, followed by Argentina (3.19 million ha) and the USA (2.2 million ha) (Figure 1).

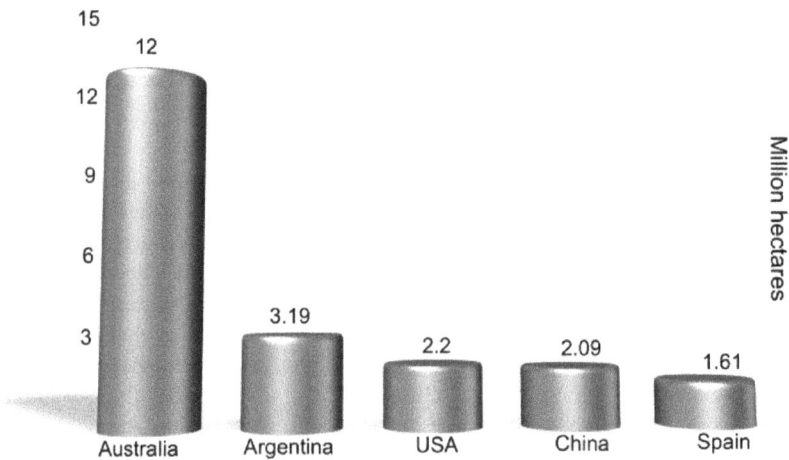

Figure 1. Countries with the largest areas of land dedicated to organic agriculture [4].

The quantity of land dedicated to organic agriculture appears to be small; however, at the local level in several countries, the impact of organic systems is very important. Although small-holder farms grow 70% of the world's food, 50% of those without food security are small-scale farmers from underdeveloped and developing countries [5]. Smallholder organic farmers from developing countries account for 73% of land certified for organic production [3]. These producers use organic techniques in soil and water and holistic management, practices that allow them to be productive, achieve food security, and increase their incomes. Ayuya et al. [6] note that organic certified smallholders are less likely to suffer multidimensional poverty compared with conventional producers.

There are an estimated two million certified organic farmers worldwide; of this total, producers in developing countries account for 80%: 34% in Africa, 29% in Asia, and 17% in Latin America [7]. The countries with the highest numbers of organic producers are India (650,000 producers), Uganda (189,610 producers), and Mexico (169,703 producers) [5] (Figure 2). Some countries, such as India, Ethiopia, Mexico, and Uganda, have promoted the participation of smallholder

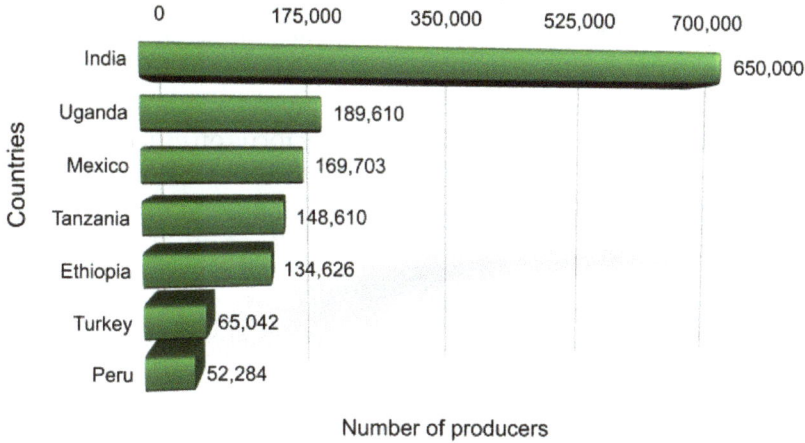

Figure 2. Countries with the largest number of certificated organic producers (adapted from [4]).

farmers in the organic market, through certification schemes such as "group certifications" and the so-called participatory guarantee systems, which link organic producers to international and domestic markets. Organic agriculture, therefore, represents an option to improve agro-ecological, social, and economic conditions in developing countries and emerging markets.

The cycle of production–consumption of certified organic products can be observed mainly in regions with high purchasing power, where consumers are able to pay the price premium of such products. In this sense, the main consuming countries of organic products are industrialized countries; the leader in organic food retail sales is the USA, with 22,590 million euros annually, followed by Germany (€7,040 million/year), France (€4,004 million/year), Canada (€2,136 million/year), and the UK (€1,950 million/year). Developed countries also have the highest consumption per capita of organic products, led by Switzerland (€189.1/year), Denmark (€165.8/year), and Luxembourg (€143.0/year) [4].

1.1. Organic livestock production

Organic livestock production is a holistic system aimed at the integration of animal and crop production and the development of a symbiotic relationship of recyclable and renewable resources [8–10]. The grassland and grazing areas used by organic livestock activity represent two-thirds (27 million hectares) of agricultural organic land; this reflects the importance of animal production within the organic production industry [4].

Organic livestock farming involves radical changes in production processes related with major attention to health and animal welfare, environmental conservation, quality, and food safety [10]. The diversity of organic livestock farms relies not only on natural local resources, animals used, climatic conditions, products manufactured, and commercialization but also on the production and farming strategies of each organic farmer.

Verhoog et al. [11] distinguish three types of organic farmers. In the "non-chemical approach," the producers are pragmatic organic farmers who formally follow organic farming standards but continue to have a conventional problem-solving approach with economic motives to conversion. The second type of producers follow the so-called "agro-ecological approach," with a more systematic approach and closed cycles; they focus on efficient production without causing damage to ecosystems. Finally, the "integrity approach" farmers develop farms where soil, plants, animals, and the farm as a whole are regarded as an organism with an intrinsic value. Each organic farming approach will determine different feed, breeding, reproduction, and health requirements.

Some of the positive effects of organic livestock practices are promoting sustainable land use, improving animal welfare and increasing product quality. The methods used exert a positive effect on biodiversity and ecological balance. Furthermore, organic management may contribute to the safeguarding of agricultural functions, with positive effects on the employment rate and the quality of life in rural communities [12, 13]. For these reasons, organic livestock farming can improve the balance between the demand for human food and the ecological impact of the agro-alimentary industry.

2. Organic dairy sheep production

Milk and dairy products constitute a high share of all organic products sales, positioned in second place behind only fruits and vegetables, and in first place for animal products, with 15% of total organic sales [14]. Sheep milk production has an important economic role in industrial countries due to high prices for dairy products, mainly cheese. Additionally, sheep milk represents a source of high quality protein and calcium in arid areas, especially for hungry or malnourished people [15].

Organic dairy sheep farms represent a system focused on producing high-quality nutritious milk, by implementing production methods that reject the use of agrochemical products, artificial compounds, pesticides, growth promoters, and forage additives and that utilize crop rotation and the reuse of organic residues. In some countries, such as Spain or Greece, organic dairy sheep systems are an essential factor for rural development for three reasons: their low environmental impact, their use of autochthonous breeds, and the diversity of transformation of milk and manufacturing processes [16].

According to Perez et al. [17], milk production is one of the most complex systems in organic production, which complicates the conversion from conventional to organic production. This is due to the large quantity of technological innovations that have been developed within the industry. However, several other authors claim that conversion from conventional to organic production systems in small ruminants appears to be less complicated in terms of management when compared with other farm species. This situation may be mainly because the management of sheep feeding does not differ dramatically between organic and conventional production systems [18–20].

Sheep have several characteristics that promote the transition process, such as easy management, effective adaptation to diverse environmental, geographic and climatic conditions, and high efficiency in the use of available sources of grazing [19, 20]. These characteristics conform with the management practices suggested by organic standards, which dictate that feeding must be based on extensive grazing and that supplementary feed should come from organic farms (certified feed industry) [21].

Organic dairy sheep farms are generally located in harsh environments, where dairy cattle production is not feasible. Organic dairy sheep nutrition is based on grazing in natural pastures and using winter fodder crops; therefore, the seasonal effects on milk sheep production are strong. The grazing system of organic dairy sheep farms promotes the continuity of traditional pastoral systems, which is the key to the sustainability of rural areas, the conservation of traditional systems of production, and the preservation of cultural heritage [22].

Location	Name of regulation	Date of publication	References
Global or regional			
FAO-WHO	*The Guidelines for the Production, Processing, Labeling and Marketing of Organically Produced Foods (Codex Guidelines)*	1999	[23]
INFOAM	*Standard for Organic Production and Processing*	August 12, 2012	[3]
EAC[1]	*East African Organic Products Standard, EAS 456:2007*	April, 2007	[24]
EU[2]	*Council Regulation (EC) No 834/2007 on organic production and labeling of organic products with regard to organic production, labeling and control*	June 28, 2007	[25]
Continent	*Country*		
America	Argentina	*National Law 25.127. Ecological, Biological and Organic Production*	September 8, 1999	[26]
	Brazil	*Law No. 10.831 and decree No. 6.323 (2007)*	December 23, 2003	[27]
	Chile	*Law 20.089 from National System of Organic Products Certification*	December 12, 2005	[28]
	Costa Rica	*Law of Development, Promotion and Foment of the Organic Agricultural activity. No. 8591*	August 14, 2007	[29]
	Mexico	*Law of Organic Products*	February 7, 2006	[30]
	United States	*National Organic Program*	December 21, 2000	[31]
Africa	Tunisia	*Law on Organic Agriculture No. 99-30*	April 5, 1999	[32]
	Uganda	*Uganda Organic Standard (UOS)* *East African Organic Products Standards*	2004 April, 2007	[24, 33]
Asia	Japan	*Japanese Agricultural Standards for Organic Livestock Products*	October 27, 2005	[34]
	India	*National Programme for Organic Production (NPOP)*	May, 2001	[35]
Oceania	Australia	*National Standard for Organic and Bio-Dynamic Produce. Edition 3.4*	July 1, 2009	[36]
	New Zealand	*Technical Rules for Organic Production. MAF Standard OP3,*	June, 2011	[37]

[1]EAC, East African Community, [2]EU, Europe Union

Table 1. Organic production standards.

The technical challenges faced by organic dairy sheep producers are regulated by international and regional standards, such as EU regulation No. 834/2007 [25], IFOAM standard for organic production and processing [3], Basic Standards and Codex Guidelines [23], and local regulations in each country (Table 1). Sheep milk production under organic management within defined standards entails challenges in feed, reproductive management, breeding, health, and welfare practices.

2.1. Feed management in organic dairy sheep farming

Organic dairy sheep systems involve extensive management, with high levels of nutrient self-sufficiency and efficient nutrient utilization. This livestock system requires management strategies with highly complex crop rotation to produce both forage and concentrate feed. Regardless of production system type (conventional or organic), the lactation process in dairy sheep requires feed rations with high levels of nutrients during mammogenesis, lactogenesis, and lactation [38]. Bencini and Pulina [39] have estimated that to produce a liter of sheep's milk with 7% fat content requires 7.1 mega joules of metabolizable energy (MJ of ME).

Country	Breed	DMY (kg/ day)	Fat %	Protein %	SNF%	TS%	References
Organic management							
Italy	Sardinian	1.23(l)	6.74	5.7	-	-	[42]
Italy	Sardinian	1.44	6.46	5.61	10.65	17.11	[43]
Czech Republic	[1]Crossbred	0.82(l)	7.94	6.49	12.25	20.19	[44]
Greece	Karagouniko	1.1	6.8	5.7	11.6	18.5	[45]
USA	[2]Crossbred	-	8.69	6.33	12.19	20.88	[46]
Czech Republic	East Friesian	1.03	6.65	5.30	11.1	17.75	[47]
Mexico	East Friesian (EF)	0.56	6.63	5.14	10.2	16.85	[48]
	EFxPelibuey	0.39	8.03	5.33	10.6	18.71	
	EFxSuffolk	0.55	6.98	5.29	10.4	17.42	
Conventional management							
Spain	Churra	1.0(l)	6.54	5.7	12.03	18.57	[49]
Israel	Awassi Assaf	2.77	4.68	5.13	-	-	[50]
Italy	Valle del Belice	1.58	7.32	5.69	-	-	[41]
Czech Republic	East Friesian	0.87	8.0	5.71	11.59	17.86	[51]
Spain	Lacaune	1.04	6.14	4.89	9.85	15.99	[52]

TMY, total milk yield; DMY, daily milk yield. SNF, Solids non-fatty; TS, Total solids, [1]First lambing crossbred ewes, Lacaune (50%), East Friesian (37.5%) and Improved Wallachian (12.5%). [2]Crossbred ewes Lacaune X East Friesian.

Table 2. Milk production and composition of dairy sheep in organic and conventional production systems.

The energy and protein content in dairy sheep rations must be adequate and sufficient to support maintenance requirements as well as milk production [40]. Pulina et al. [41] note that energy intake is the most important factor that influences milk production and composition, followed by protein and fiber content of the diet. An adequate amount of energy in dairy sheep diets increases glucose content in the blood, which promotes the synthesis of lactose, the activation of mammary and systemic regulators (insulin, IGF, thyroid and neurohormones, etc.), and the increased uptake of milk precursors (glucose, acetate, butyrate, amino acids, NEFA, vitamins, and minerals) [41].

The standards that regulate feeding management in organic systems are one of the most critical factors that influence milk production performance and quality of milk (Table 2). Organic regulations limit the use of concentrate and reduce the range of ingredients that can be included in organic rations. This situation may cause deficiencies of energy, protein, and minerals (zinc, molybdenum, selenium, copper, and iodine), which increases the risk of nutrient imbalances [53, 54]; it has been reported that underfeeding ewes in early lactation, when nutritional requirements are highest, results in lower milk yields [55].

European organic standards require feed rations based on forage (minimum 60% of daily dry matter inclusion) and primarily homegrown ingredients [25]. One of the major challenges in organic management is to formulate high forage diets with an adequate energy concentration due to the low energy value of most forages (<11 MJ of ME per kg DM) when compared to concentrate feeds (>13 MJ of ME per kg DM) [56]. The stage of lactation determines the percentage of forage in the total ration, which can comprise up to 100% of the total ration. Organic dairy sheep can graze in natural or cultivated pastures, and different strategies of feeding can be used to follow organic standards.

The feed management on most organic dairy sheep farms is based on grazing. Grazing is the interaction between animals using the pasture and the pasture itself [57]. Systems based on natural pasture grazing utilize less fertilizer and are considered more ecological. However, the high level of pasture in diet, the availability and quality of forage, and the change from grazing fresh herbage to consuming conserved forage are associated with lower milk yields for sheep under organic management compared with milk yields on conventional farms [58, 59].

The availability and quality of pastures and conserved forage change significantly throughout the year, producing a seasonality effect on milk production. Angeles-Hernandez et al. [60] analyze the effect of lambing season on milk production in sheep under organic management; they conclude that sheep with autumn lambing showed significantly ($P = 0.002$) higher milk yields (Figure 3). This may be due to the sheep having been pregnant during the summer, when the availability of forage reaches its maximum, producing a positive effect on the differentiation of mammary secretory cells as well as on the buildup of the animal's physical condition.

Zervas et al. [58] analyze the milk production and live-weight changes in ewes in both conventional and organic systems. Ewes under organic management were fed with grass hay plus barley grain, and ewes under conventional management were fed with grass hay plus balanced concentrate feed. Milk yields of ewes fed organically were significantly lower ($P <$

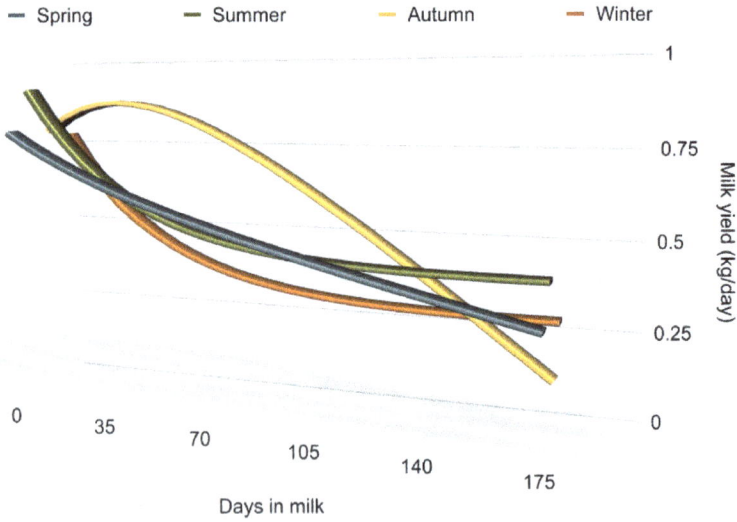

Lactation curve adjusted using the Wilmink model [61] ($Y = a + be^{kt} + ct.$)

Figure 3. Lactation curves per lambing season of dairy sheep under organic management (adapted from [60]).

0.001) when compared with conventional-fed ewes (134 vs. 180 kg/year, respectively). Also, ewes in conventional management showed higher values of live-weight gain ($P < 0.01$) in the period between lambing and weaning (organic 67 vs. conventional 79 g/day).

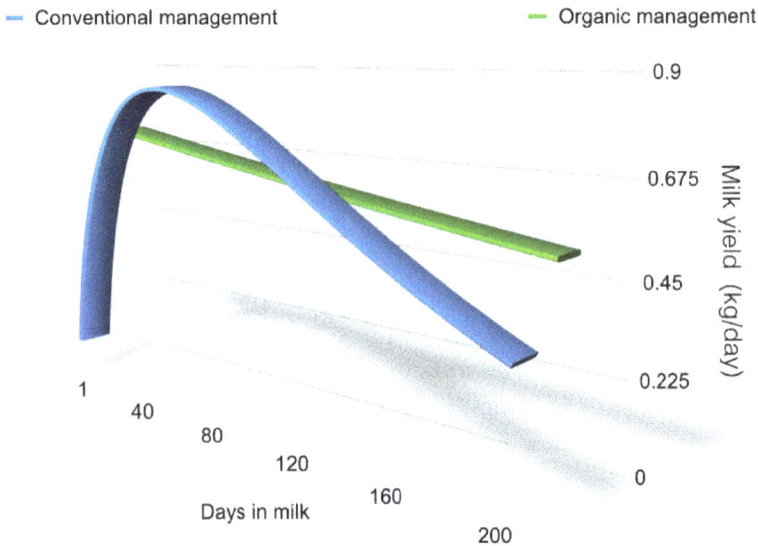

Figure 4. Lactation curves of conventional and organic dairy sheep farms (adapted from [62]).

Some studies note that milk yields of dairy sheep under organic management can be similar or higher than conventional dairy farms, which can be explained in part by lower stocking rates and high availability of forage per animal [43, 63]. Angeles-Hernandez and Gonzalez-Ronquillo [62] compared the milk production and lactation curves of conventional and organic dairy sheep farms; these authors used the Wood model [64] to analyze a total of 7,501 weekly test-day milk yield records from crossbred dairy ewes. There were no differences in milk yields between organic and conventional dairy sheep farms (97 vs. 103 kg, respectively), but there were significant differences ($P < 0.05$) in the shape of the lactation curve (Figure 4), traits that defined the shape of lactation curve (peak yield and time of peak yield), and parameters of the Wood model (Table 3). Sheep in organic systems showed a higher percentage of lactation curves with atypical shape (without peak of lactation) (Table 3), which could be beneficial in this system, as the risk of negative energy balance and metabolic disturbances in early lactation is lower (Figure 4).

Type of farming	Traits of lactation curve			Parameters of Wood model			Proportion of atypical shapes
	TMY(kg)[1]	PY(kg)	PT(kg)	a	b	c	(%)
Organic	97.3	0.79[b]	20.9[b]	0.51[a]	0.43[b]	0.011[a]	52.6
Conventional	103.0	0.85[a]	86.9[a]	0.25[b]	1.89[a]	0.002[b]	10.5
P-value	0.06	0.05	0.001	0.01	0.001	0.001	

[1] TMY, total milk yield adjusted to 200 days in milk; PY, peak yield; PT, time to peak yield; a is the production of milk at beginning of the lactation (kg), b and c are parameters of inclining and declining slopes of lactation curve before and after the PY, respectively.

Table 3. Characteristics of lactation curve and parameters of Wood model from lactation of organic and conventional dairy sheep farming (Adapted from [62]).

Pasture farming systems result in milk characterized by a chemical composition that has beneficial properties for human health. Organic sheep milk has a high fat content (Table 2) due to rations rich in fiber [15]. Several studies report that milk and dairy products from certified organic production systems contain higher concentrations of protein, cis-9, trans-11 CLA, α-linolenic (α-LNA), transvaccenic acid, docosapentanoic acid, eicosapentanoic acid, total n-3 fatty acids, α-tocopherol, and β-carotene than those from conventional production systems [65–67]. Tsiplakou et al. [45] conclude that sheep milk produced under organic farming conditions has higher nutritive values, with elevated contents of MUFA, PUFA, α-LNA, cis-9, trans-11 CLA, and ω-3 FA compared with that from conventional systems.

2.2. Effect of genetic factors in organic dairy sheep farming

The breed or genotype of dairy sheep is one of the main factors that affects milk yields and chemical composition. The choice of breed in organic systems must be considered, with an emphasis on animal characteristics that ensure their welfare and health, such as adaptation to local environmental conditions, vitality and resistance to disease, and absence of specific health problems associated with certain breeds [23, 25].

According to Nauta et al. [2], the different production and marketing strategies of organic farmers demand different breeds. Current dairy breeds have been modified through selective breeding programs to produce high levels of milk, which may make them unsuitable for a traditional and more natural production system. However, the "non-chemical approach" organic farmers use specialized dairy sheep breeds to reach economically viable milk yields, and organic farmers with other production approaches use specialized dairy sheep breeds during the conversion process, usually with moderate milk production performance (Table 2).

The main strategies of animal breeding in organic dairy systems are selection (within and among breeds) and crossbreeding. Selection in organic farming should be used to reinforce, in a sustainable manner, the relationship between the animal and the environment in which it is produced [21]. There are differences in the characteristics and magnitude of genotype due to external factors (i.e., environmental interaction between conventional and organic systems) [68]; the specific approximation to environmental conditions of organic management determines different selection traits for both production systems (Table 4).

The program of selection on organic dairy sheep farms can be applied to specialized, local, or native breeds to improve dairy production traits, but it mainly promotes the selection of vital traits that improve animal well-being, sustainability, health, and flock efficiency [69] (Table 4). Nauta et al. [2] noted that 43% of organic farmers were seeking functional traits as a breeding goal, 32% productive traits, and 25% conformation traits.

Trait	Heritability
General disease resistance	0.05-0.80
Resistance to parasite infection	0.25-0.40
Somatic cell count	0.12-0.13
Longevity	0.05-0.13
Female fertility	0.07-0.20
Mature size	0.47
Feeding characteristics	0.10
Udder shape	0.20-0.24
Teat size	0.18-0.39
Milking ease	0.01
Milk production and composition	
Milk production	0.28-0.32
Fat content	0.41-0.62
Protein content	0.51-0.53
Fat yield	0.17-0.29
Protein yield	0.18-0.27

Data from: [21, 70-77].

Table 4. Important traits in organic dairy sheep breeding.

Organic dairy production can benefit from using native or local breeds genetically adapted to their environment; these breeds are more resilient to climatic stress and are resistant to local parasites and diseases, enabling them to utilize a lower quality of feed [78]. Organic farming may contribute to the maintenance and improvement of the variability of dairy sheep breeds. The use of native breeds can also help support food, agricultural, and cultural diversity, in that the milk and cheese produced from sheep are an expression of a regional cultural tradition. Native breeds also promote local food security and represent a valuable genetic source for improving health and performance traits in the future [12, 78]. However, under organic management, the use of local sheep breeds that are not specialized in milk production may hinder the achievement of sufficient milk yields to reach economic viability. In these situations, crossbreeding can be an option as an improved genetic strategy [79].

Crossbreeding of native breeds with specialized dairy breeds is a viable option to improve dairy production parameters and promote adaptation to feed sources, climate, and the management and market conditions of organic milk production systems, through heterosis and the combined attributes of different breeds [48]. When animals are genetically adapted to specific/extreme environmental conditions, they will be more productive and production costs will be lower. Furthermore, genetic groups adapted to organic dairy management help to safeguard animal health and welfare [78].

Angeles-Hernandez et al. [48] carried out an evaluation of the effect of genetic group on milk production and composition on an organic dairy sheep farm; they compared three genetic groups: East Friesian (EF), EFxPelibuey (local hair breed) (EFxPL), and EFxSuffolk (EFxSF). They found significant differences among genetic groups in milk yield and milk composition (Figure 5). EFxPL sheep showed a lower milk yield (59.8 kg), protein yield (20.8 g/day), and fat yield (31.3 g/day) compared to the other groups. EF and EFxSF showed similar values of milk yield (76.1 $\cong$ 75.8 kg), protein yield (28.8 $\cong$ 29.1 g/day), and fat yield (37.2 $\cong$ 38.4 g/day, respectively).

The EFxSF group showed appropriate milk yield and chemical composition; these contribute not only to an increased cheese yield but also to a differentiation of cheese flavor. However, crossbreeding presents challenges in terms of maintaining a correct proportion of purebred–crossbred populations; furthermore, in systems with inadequate management, biodiversity may be jeopardized by the elimination of certain purebreds (specialized and native breeds) [21].

The goals of organic dairy production farms are more than maximum milk productivity; their objectives are directed to favoring animal health and welfare and to improving the quality of their products with minimum environmental impact. In this sense, genetic improvement strategies must be individually selected and designed for each farm according to resource availability, local market conditions, and management approach.

2.3. Economic implications of organic dairy sheep farming

Organic dairy sheep farming provides income to thousands of families and contributes to regional development, especially in isolated and less favored areas. It also generates employ-

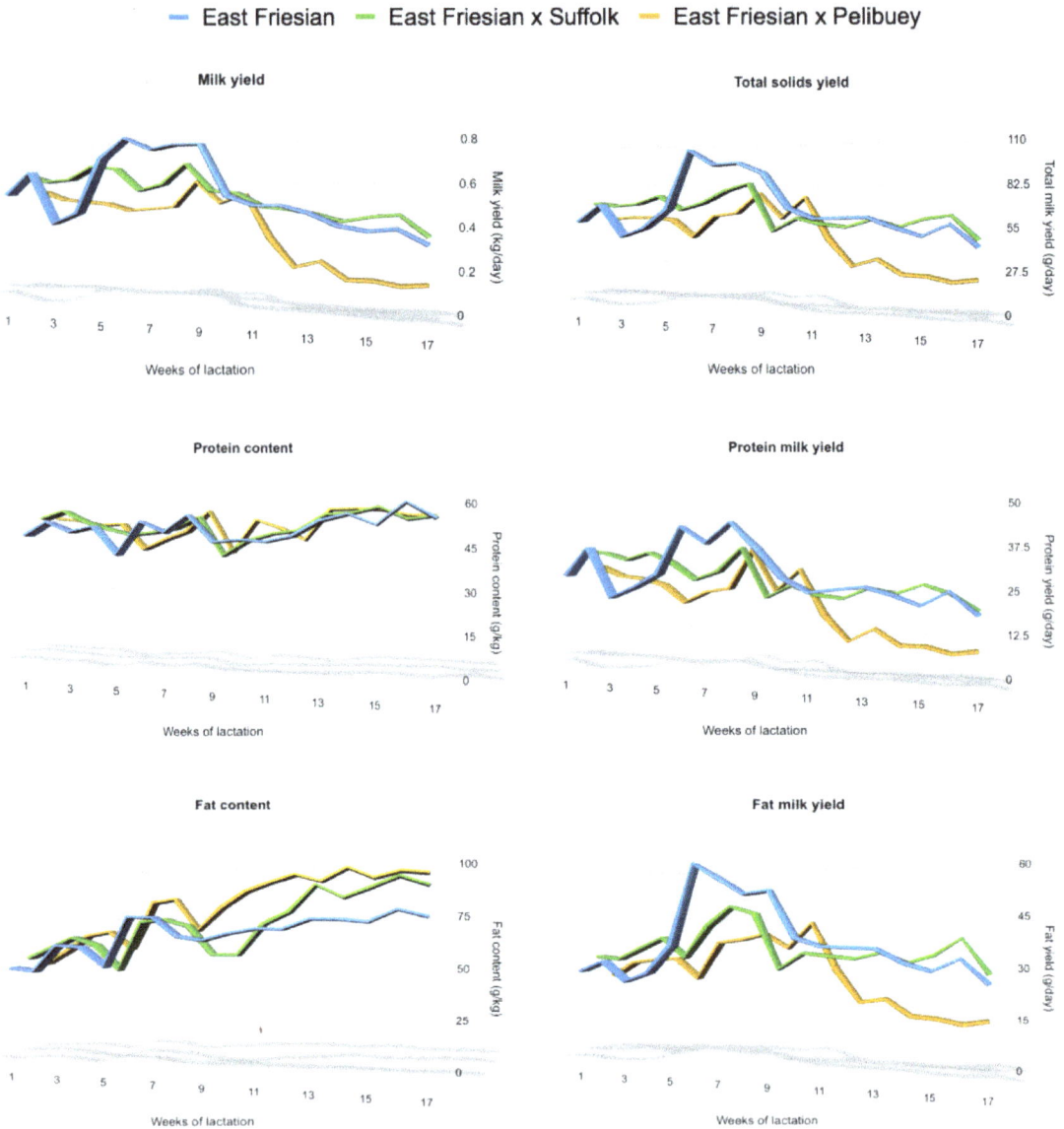

Figure 5. Effect of genetic group on milk production and composition in sheep under organic management (adapted from [48]).

ment, promotes closer links with local markets, restores connections between farmers and customers, and increases incomes in the local economy through exports [13].

The specific productive approach of organic dairy sheep farms determines its economic stability and profitability. The main factors that affect the expected returns of dairy sheep farming are milk yield and price of dairy products [80]. The competitive prices of organic products has played an important role in the expansion of interest in organic systems. Frequently, organic products obtain a premium price when compared to products from conventional farms. The magnitude of the premium depends on product availability and market demand.

The premium in price for organic sheep milk over conventional milk ranges from 8% to 36% within European market [81], 51% in New Zealand [82], from 47% to 79% in the USA [83], and a price difference of approximately 20% to 30% in Mexico [79]. In the case of the gross production value of meat and lambs, the variation arises mainly from fluctuations in price. Gross production value for ewe meat (non-productive ewes) contributes less to the total gross production value of the farm.

Gerrard et al. [84] have noted that organic dairy sheep farms show lower investments in items such as acquisition of animals, equipment depreciation, and agronomic management (less use of fertilizers and chemical compounds). However, it should be taken into account that in the case of organic farming, the value of animal capital is lower due to the fact that the flock consists mainly of crossbred dairy ewes [79]. It has also been reported that organic dairy sheep farms employ more people in comparison with conventional farms. Padel and Lampink [85] noted the higher number of working hours on organic farms (10–50% greater), and they considered salaries to be an expense with a higher impact on the total cost of organic milk sheep production.

The initial investment for establishing an organic farm, as for a conventional farm, includes investments in buildings (stables and barns), equipment (milking machine, feeders), animal capital, pasture area, and grain supplements for feeding throughout the year. An added investment that needs to be considered for organic farms is the certification process, as well as the fact that during the conversion process the commercialization of dairy products with a premium price is not yet possible.

The questions that we have to ask when comparing conventional systems vs. organic systems in general terms are as follow: How will sustainable intensification work in practice? How can farmers and other producers improve their production systems to produce food in more sustainable ways? Being less susceptible to volatile food prices, how can niche-level innovations and consumer interpretations and social practices be better integrated into the mainstream food security discourse? For example, organic systems offer the security of avoiding chemical fertilizers, antibiotics, hormones, and synthetic growth promoters, all of which involve human risk through the increase in allergies and antimicrobial resistance. How will the transformations of the food system play out in terms of geographical area, food security and animal welfare?

From the economic perspective, the dominant message is the importance of the profit motive, which drives the production system. However, the cost to the environment must also be taken into account. For this reason, we have to analyze the "economic sustainability" based not only on economic profitability but also on the relationship of farmers to their land environment and the sustainability of their activity [86]. There may also be hidden costs of production not only from agricultural intensification [87] but also from organic production [88].

3. Conclusion

Organic production is not a method of production that can solve all the problems of the dairy sheep industry; it is mainly an approach to production focused on satisfying the current demand for dairy products, but without the adverse effects of intensified livestock production. Moreover, organic farming is a production method with a specific market focus on products of premium quality and high standards of production. Organic sheep milk production can provide a balance between society's demand for food and the ecological impact of the agro-alimentary industry, through the comprehensive implementation of conservation practices and the ecological utilization of natural resources.

The production of organic sheep's milk requires research along specific lines, aimed at developing better methods of production, distribution, and marketing of their products. These must be focused mainly on genetic improvement, preventive medicine, welfare, nutrition management, and promotion of nutritional characteristics, in accordance with defined production approaches and regulations.

Acknowledgements

Mr. Angeles Hernandez was granted with CONACyT, Mexico scholarship for their studies in the Universidad Nacional Autonoma de Mexico. Dr. Gonzalez Ronquillo was granted with a CONACyT fellowship "Estancias Sabaticas en el Extranjero 2014." Also, we thank Ms. Penelope Krumm for the critical review of this paper.

Author details

Juan C. Angeles Hernandez[1,4], Octavio A. Castelan Ortega[2], Sergio Radic Schilling[3], Sergio Angeles Campos[4], A. Hilda Ramirez Perez[4] and Manuel Gonzalez Ronquillo[2*]

*Address all correspondence to: mrg@uaemex.mx

1 Programa de Maestría y Doctorado en Ciencias de la Producción y de la Salud Animal, Facultad de Medicina Veterinaria y Zootecnia, Universidad Nacional Autónoma de México, Circuito Exterior, Ciudad Universitaria, Delegación Coyoacán, México

2 Universidad Autónoma del Estado de México, Instituto Literario 100 Ote, Toluca, Estado de México, México

3 Escuela de Ciencias y Tecnologías en Recursos Agrícolas y Acuícolas, Universidad de Magallanes, Región de Magallanes y la Antártica Chilena, Chile

4 Facultad de Medicina Veterinaria y Zootecnia, Departamento de Nutrición Animal y Bioquímica, Universidad Nacional Autónoma de México, Circuito Exterior, Ciudad Universitaria, Delegación Coyoacán, México

References

[1] Terrence T, Cihat G. Organic agriculture, sustainability and consumer preferences. In: Vytautas P, editor. Organic Agriculture Towards Sustainability. InTech; 2014. p. 1–24. DOI: 10.5772/58428

[2] Nauta WJ, Baars T, Bovenhuis H. Converting to organic dairy farming: consequences for production, somatic cell scores and calving interval of first parity Holstein cows. Livestock Science. 2006;90:185–195. DOI: 10.1016/j.livprodsci.2005.06.013

[3] International Federation of Organic Agriculture Movements, IFOAM. The IFOAM standard for organic production and processing; 2012. Available from: http://www.ifoam.bio/en/ifoam-standard [Accessed: 2014-12-20]

[4] FiBL-INFOAM. The World of Organic Agriculture 2015. Available from: https://www.fibl.org/fileadmin/documents/shop/1663-organic-world-2015.pdf. [Accessed: 2015-01-12]

[5] INFOAM. Consolidated Annual Report of the IFOAM Action Group 2013. Available from: http://www.ifoam.bio/sites/default/files/annual_report_2013_web.pdf. [Accessed: 2015-02-21]

[6] Ayuya OI, Gido EO, Bett HK, Lagat JK, Kahi AK, Bauer S. Effect of certified organic production systems on poverty among smallholder farmers: empirical evidence from Kenya. World Development. 2015;67:27–37. DOI: 10.1016/j.worlddev.2014.10.005

[7] Global Organic Market Access GOMA. Conference to highlight 10 years of growing market access for organic agriculture. In: Proceedings of the Global Organic Market Access Conference. February 2012, Nuremberg, Germany; 2002. p. 39–40.

[8] Alrøe HF, Kristensen ES, Halberg, N. A systems approach to research in sustainability and organic famring. In: Research methodologies in organic farming. Workshop in Frink. September–October 1998; 30-FAO REU Technical series 58.

[9] Siardos GC. The impact of organic agriculture in socio-economic structures. In: Kyriazakis I, Zervas G, editors. Organic Meat and Milk from Ruminants. Netherlands: Wageningen Academic Publishers; 2002. p. 73–86. DOI: 10.3920/978-90-8686-506-2

[10] Blair, R. Aims and principles of organic cattle production. In: Blair R, editor. Nutrition and Feeding of Organic Cattle. UK: CAB International; 2011. p. 5–27. DOI: 10.1079/9781845937584.0000

[11] Verhoog H, Matze M, Lammerts Van Bueren E, Baars T. The role of the concept of the natural (naturalness) in organic farming. Journal of Agricultural and Environmental Ethics. 2003;16:29–49. DOI: 10.1023/A:1021714632012

[12] Ronchi B, Nardone A. Contribution of organic farming to increase sustainability of Mediterranean small ruminant livestock systems. Livestock Production Science. 2003;80:17–31. DOI: 10.1016/S0301-6226(02)00316-0

[13] Lobley M, Butler A, Winter M. Local organic food for local people? Organic marketing strategies in England and Wales. Regional Studies. 2013;47(2): 216–228. DOI: 10.1080/00343404.2010.546780

[14] Johnson J. Saving natural. Nutrition Business Journal. 2015: 3–11. Available from: http://newhope360.com/sitefiles/newhope360.com/files/uploads/2015/02/2015%20nbjSampleIssue_CZ.pdf [Accessed: 2015-01-25]

[15] Morand-Fehr P, Fedele V, Decandia M, Le Frileux Y. Influence of farming and feeding systems on composition and quality of goat and sheep milk. Small Ruminant Research. 2007;68(1):20–34. DOI: 10.1016/j.smallrumres.2006.09.019

[16] Toro-Mujica PM. Análisis técnico, económico y social de los sistemas ovino lechero ecológicos de Castilla la Mancha: Eficiencia y sustentabilidad [thesis]. Spain: Córdoba University; 2011 (In Spanish).

[17] Pérez BL, Esquivel CG, Molina OM, Hernández AH. Importancia económica y normatividad de la producción de leche orgánica. In: Hernandez LAG, Pérez LB, editors. Producción sustentable calidad y leche orgánica. México: Universidad Autónoma Metropolitana; 2009. p. 85–105 (In Spanish).

[18] Wright IA, Zervas G, Louloudis L. The development of sustainable farming systems and the challenges that face producers in the EU. In: Kyriazakis I, Zervas G, editors. Organic Meat and Milk from Ruminants. Netherlands: Wageningen Academic Publishers; 2002. p. 27–37.

[19] Degen AA. Sheep and goat milk in pastoral societies. Small Ruminant Research. 2007;68:7–19. DOI: 10.1016/j.smallrumres.2006.09.020

[20] Shrestha JNB. Sheep. In: Fuquay JW, Fox PF, McSweeney PLH, editors. Encyclopedia of Dairy Science. Elsevier. UK: CABI Publishing; 2011. p. 67–76.

[21] Nardone A, Zervas G, Rochi B. Sustainability of small ruminant organic systems of production. Livestock Production Science. 2004;90:27–39. DOI: 10.1016/j.livprodsci.2004.07.004

[22] Paniagua A. The politics of place: official, intermediate and community discourses in depopulated rural areas of Central Spain. The case of the Riaza river valley (Segovia, Spain). Journal of Rural Studies. 2009;25(2):207–216. DOI: 10.1016/j.jrurstud.2008.12.001

[23] FAO/WHO Food Standards Programme-Codex Alimentarius Commission. Guidelines for the production, processing, labeling and marketing of organically produced foods. 1999. Available from: http://www.fao.org/organicag/doc/glorganicfinal.pdf [Accessed: 2015-02-19]

[24] East African Community EAC. East African Organic Products standards EAS456:2007; 2007. Available from: https://www.organic-standards.info/en/documents/East-African-Organic-Product-standard,25. [Accessed: 2015-02-20]

[25] European Union (EU). Council Regulation (EC) No. 834/2007 on organic production and labeling of organic products with regard to organic production, labeling and control. 2007. Official Journal of the European Union. Available from: http://eurlex.europa.eu/legalcontent/EN/TXT/?qid=1407328133792&uri=CELEX: 32007R0834. [Accessed: 2015-01-15]

[26] Servicio Nacional de Sanidad y Calidad Agroalimentaria de Argentina SENASA. Produccion Ecologica, Biologica y Organica, Ley 25.127. 1999. Available from: http://www.senasa.gov.ar/contenido.php?to=n&in=981&io=4659. [Accessed: 2015-01-22] (In Spanish).

[27] Ministro de estado da agricultura, pecuária e abastecimento MAPA. lei 10.831. Diário Oficial da União. 2003. Available from: http://www.agricultura.gov.br/arq_editor/file/Desenvolvimento_Sustentavel/Organicos/Produtos%20Fitossanitários/Home/Lei_10-831_de_2003 (portuguese).pdf [Accessed: 2015-01-19].

[28] Servicio Agricola y Ganadero SAG. Law No. 20.089. (2005). Available from http://www.sag.cl/sites/default/files/Ley_reglamento_version_dic2011.pdf   [Accessed: 2015-01-27] (In Portuguese)

[29] Ministerio de Agricultura y Ganaderia-Hacienda-Economia, Industria y Comercio, MAG-H-MEIC. Desarrollo, promoción y fomento de la actividad agropecuaria orgánica. No. 8591. 2007. Available from: http://www.mag.go.cr/legislacion/2009/de-35242.pdf [Accessed: 2015-01-19] (In Spanish).

[30] Secretaría de Agricultura, Ganadería, Desarrollo Rural, Pesca y Alimentación SAGARPA. Law of Organic Products. 2006. Available from: http://www.sagarpa.gob.mx/normateca/Normateca/2%20DECRETO%20por%20el%20que%20se%20expide%20la%20Ley%20de%20Productos%20Orgánicos%20070206.pdf.   [Accessed: 2015-01-25] (In Spanish).

[31] USDA. National Organic Program. Part 205. 2000. Available from: http://www.ecfr.gov/cgibin/textidx?SID=b7698166035019cd62a11e2d5be520ba&mc=true&node=pt7.3.205&rgn=div5 [Accessed: 2015-02-11].

[32] Journal Officiel de la Republique Tunisienne JORT. Law on Organic Agriculture No. 99-30; 1999. p. 539–540. Available from: http://faolex.fao.org/docs/pdf/tun16328.pdf [Accessed: 2015-01-21].

[33] National Organic Agricultural Movement of Uganda NOGAMU and Uganda National Bureau of Standards UNBS. Uganda organic standard (UOS) for organic production and processing. 2004. Available from: http://www.grolink.se/epopa/Publications/UOS_July-05.pdf [Accessed: 2015-02-19].

[34] Ministry of Agriculture, Forestry and Fisheries MIFF. Japanese Agricultural Standard for Organic Livestock Products Established: Notification No.59. (2005). Available

from: http://www.maff.go.jp/e/jas/specific/pdf/836_2012-2.pdf   [Accessed: 2015-03-27].

[35] Ministry of Commerce and Industry, Government of India MCI. National Programme for Organic Production (NPOP). 2001. Available from: http://apeda.gov.in/apedawebsite/organic/ORGANIC_CONTENTS/National_Programme_for_Organic_Production.htm [Accessed: 2015-02-17].

[36] National Standard for Organic and Bio-Dynamic Produce. Organic Industry Export Consultative Committee-Australian Quarantine and Inspection Service (OIECC-AQIS). 2009. Available from: http://www.tamburlaine.com.au/docs/national-standard.pdf [Accessed: 2015-02-01].

[37] Ministry of Agriculture and Forestry of New Zealand MAF. Technical Rules for Organic Production MAF Standard OP3. 2011. Available from: http://www.foodsafety.govt.nz/elibrary/industry/nzfsa-standard-registration-documents/index.htm [Accessed: 201-12-15].

[38] Tucker HA. Physiological Control of Mammary Growth, Lactogenesis, and Lactation. Journal of Dairy Science. 1981; 64(6):1403–1421.

[39] Bencini R, Pulina G. The quality of sheep milk: a review. Animal Production Science. 1997; 37(4):485–504. DOI: 10.1071/EA96014

[40] Gilbert G. Nutrition of dairy sheep. In: Dawe ST, Dignand M, editors. Sheep Dairying: The Manual. NSW Agriculture; 1992. p. 1–5.

[41] Pulina G, Macciotta NPP, Nudda A. Milk composition and feeding in the Italian dairy sheep. Italian Journal of Animal Science. 2005;4(Suppl. 1):5–14. DOI: 10.4081/ijas.2005.631

[42] Ronchi B, Bernabucci U, Amici A, Serrani F. Effects of different source of yeast on milk yield and composition in organic dairy sheep farming. In: Kyriazakis I, Zervas G editors. Organic Meat and Milk from Ruminants. Netherlands: Wageningen Academic Publishers; 2002. p. 189–193. DOI: 10.3920/978-90-8686-506-2

[43] Pirisi A, Piredda G, Sitzia M, Fois N. Organic and conventional systems: composition and chesse-aking aptitude of Sarda ewes´s milk. In: Kyriazakis I, Zervas G, editors. Organic Meat and Milk from Ruminants. Netherlands: Wageningen Academic Publishers; 2002. p. 143–146. DOI: 10.3920/978-90-8686-506-2

[44] Novotná L, Kuchtík J, Ustová K, Zapletal D, Filip [cbreve] ík R. Effects of lactation stage and parity on milk yield, composition and properties of organic sheep milk. Journal   of   Applied   Animal   Research.   2009;36(1):71–76.   DOI: 10.1080/09712119.2009.9707034

[45] Tsiplakou E, Kotrotsios V, Hadjigeorgiou I, Zervas G. Differences in sheep and goats milk fatty acid profile between conventional and organic farming systems. Journal of Dairy Research. 2010; 77:343–349. DOI: 10.1017/S0022029910000270

[46] King R. Managing dairy sheep nutrition on a organic farm. In: Proceeding of the 18th Annual Dairy Sheep Association of North America Symposium; 18–20 October 2012; Virginia, USA. p. 22–26.

[47] Králíčková Š, Pokorná M, Kuchtík J, Filipčík R. Effect of parity and stage of lactation on milk yield, composition and quality of organic sheep milk. Acta Universitatis Agriculturae et Silviculturae Mendelianae Brunensis. 2013;60(1):71–78. DOI: 10.11118/actaun201260010071

[48] Angeles-Hernandez JC, Castelan OOA, Ramirez-Perez AH, González-Ronquillo M. Effects of crossbreeding on milk production and composition in dairy sheep under organic management. Animal Production Science. 2014;54:1641–1645. DOI: 10.1071/AN14214

[49] Fuertes JA, Gonzalo C, Carriedo JA, San Primitivo F. Parameters of Test Day Milk Yield and Milk Components for Dairy Ewes. Journal of Dairy Science. 1998;81(5): 1300–1307. DOI: 10.3168/jds.S0022-0302(98)75692-9

[50] Leitner G, Chaffer M, Caraso Y, Ezra E, Kababea D, Winkler M, Saran A. Udder infection and milk somatic cell count, NAGase activity and milk composition—fat, protein and lactose—in Israeli-Assaf and Awassi sheep. Small Ruminant Research. 2003;49(2):157–164. DOI: 10.1016/S0921-4488(03)00079-8

[51] Kuchtík J, Sustova K, Urban T, Zapletal D. Effect of the stage of lactation on milk composition, its properties and the quality of rennet curdling in East Friesian ewes. Czech Journal of Animal Science. 2008;53(2):55.

[52] Andrade BR, Salama AAK, Caja G, Castillo V, Albanell E, Such X. Response to lactation induction differs by season of year and breed of dairy ewes. Journal of Dairy Science. 2008;91(6):2299–2306. DOI: 10.3168/jds.2007-0687

[53] Coonan C, Freestone-Smith C, Allen J, Wilde, D. Determination of the major mineral and trace element balance of dairy cows in organic production systems. In: Kyriazakis I, Zervas G, editors. Organic Meat and Milk from Ruminants. Netherlands: Wageningen Academic Publishers; 2002. p. 181– 183. DOI: 10.3920/978-90-8686-506-2

[54] Rosati A, Aumaitre A. Organic dairy farming in Europe. Livestock Production Science. 2004;90(1):41–51. DOI: 10.1016/j.livprodsci.2004.07.005

[55] Byström S, Jonsson S, Martinsson K. Organic versus conventional dairy farming– studies from the Öjebyn Project. In: Proceedings of the UK Organic Research 2002 Conference. Aberystwyth: Organic Centre Wales, Institute of Rural Studies, University of Wales. p. 179–184

[56] Weller, R. Improving organic milk. In: Griffiths MW, editor. Improving the Safety and Quality of Milk. Volume 2: Improving Quality in Milk Products. Wood Publishing Series in Food Science, Technology and Nutrition; 2010. p. 283–303.

[57] Molle G, Decandia M, Ligios S, Fois N, Treacher TT, Sitzia M. Grazing Management and stocking rate with particular reference to the Mediterranean environment. In:

Pulina G, Bencini R, editors. Dairy Sheep Nutrition. CABI Publishing; 2004. p. 191–211.

[58] Zervas G, Koutsotolis K, Theodoropoulos G, Zabeli G, Gagnaux D, Poffet JR. Comparison of organic with conventional feeding systems of lactating dairy ewes in Greece. In: Proceedings of the Fifth International Symposium on Livestock Farming Systems; 19–20 August, 1999; Posieux, Fribourg, Switzerland: Wageningen Pers; 2000. p. 107–111.

[59] Von Borell E, Sørensen JT. Organic livestock production in Europe: aims, rules and trends with special emphasis on animal health and welfare. Livestock Production Science. 2004;90(1):3–9.

[60] Angeles-Hernandez JC, Albarran-Portillo B, Gonzalez AG, Salas NP, Gonzalez-Ronquillo M. Comparison of mathematical models applied to f1 dairy sheep lactations in organic farm and environmental factors affecting lactation curve parameter. Asian-Australasian Journal of Animal Sciences. 2013;26(8):1119–1126. DOI: 10.5713%2Fajas.2013.13096

[61] Wilmink JBM. Adjustment of test-day milk, fat and protein yield for age, season and stage of lactation. Livestock Production Science. 1987;16:335–348. DOI: 10.1016/0301-6226(87)90003-0

[62] Angeles-Hernandez JC, González-Ronquillo M. Comparing the milk production and characteristics of lactation curves in organic and conventional dairy sheep farms in Mexico. In: Proceeding of International Symposium on Food Safety and Quality: Applications of Nuclear and Related Techniques, 10–13 November 2014. Vienna. IAEA; 2006.

[63] Padel S. Strategies of organic milk production. In: Proceedings of the 3th NAHWOA Workshop, Clermont-Ferrand, France; 2000. p. 121–135.

[64] Wood PDP. Algebraic model of the lactation curve in cattle. Nature. 1967;216:164–165. DOI: 10.1038/216164a0

[65] Bisig W, Eberhard P, Collomb M, Rehberger B. Influence of processing on the fatty acid composition and the content of conjugated linoleic acid in organic and conventional dairy products—a review. Le Lait. 2007;87:1–19. DOI: 10.1051/lait:2007001

[66] Palupi E, Jayanegara A, Ploeger AJ, Kahl J. Comparison of nutritional quality between conventional and organic dairy products: a meta-analysis. Journal of the Science of Food and Agriculture. 2012;92:2774–2781. DOI: 10.1002/jsfa.5639.

[67] Heins BJ, Paulson JC, Endres MI, Moon RD. Effect of organic grain supplementation on pasture and total mixed ration dry matter intake and fatty acid profiles of organic dairy cows. In: Join Annual Meeting ADSA-ASAS. 8–14 July 2013. Indianapolis. p. 662.

[68] Nauta WJ. Selective Breeding in Organic Dairy Production [thesis]. Wageningen, the Netherlands: Wageningen University; 2009.

[69] Conington J, Lewis RM, Simm G. Breeding goals and strategies for organic sheep production. In: Kyriazakis I, Zervas G, editors. Organic Meat and Milk from Ruminants. Netherlands: Wageningen Academic Publishers; 2002. p. 135– 142. DOI: 10.3920/978-90-8686-506-2

[70] Mavrogenis AP, Papachristoforou C, Lysandrides P, Roushias A. Environmental and genetic factors affecting udder characters and milk production in Chios sheep. Genetics Selection Evolution. 1988;20:477–488. DOI: 10.1186/1297-9686-20-4-477

[71] Fernández G, Baro JA, De la Fuente LF, San Primitivo F. Genetic parameters for linear udder traits of dairy ewes. Journal of Dairy Science. 1997;80:601–605. DOI: 10.3168/jds.S0022-0302(97)75976-9

[72] Conington J, Bishop SC, Grundy B, Waterhouse A, Simm G. Multi-trait selection indexes for sustainable UK hill sheep production. Animal Science. 2001;73:413–424.

[73] Rupp R, Lagriffoul G, Astruc JM, Barillet F. Genetic parameters for milk somatic cell scores and relationships with production traits in French Lacaune dairy sheep. Journal of Dairy Science. 2003;86(4):1476–1481. DOI: 10.3168/jds.S0022-0302(03)73732-1

[74] Berger Y, Billon P, Bocquier F, Caja G, Cannas A, McKusick B, Marnet PG, Thomas D. Principles of sheep dairying in North America. 2004. Available from: http://www.drobnica.si/files/brosure/Berger_et_al_Principles_of_Sheep_Dairying_in_North_America_2004.pdf [Accessed: 2015-02-17]

[75] Legarra A, Ugarte E. Genetic parameters of udder traits, somatic cell score, and milk yield in latxa sheep. Journal of Dairy Science. 2005;88:2238–2245. DOI: 10.3168/jds.S0022-0302(05)72899-X

[76] Legarra A, Ramón M, Ugarte E, Pérez-Guzmán MD. Economic weights of fertility, prolificacy, milk yield and longevity in dairy sheep. Animal. 2007;01:193–203. DOI: 10.1017/S1751731107657814

[77] Ugarte E. The breeding program of Latxa breed. Biotechnology in Animal Husbandry. 2007;23:97–111.

[78] Van Diepen P, McLean B, Frost D. Livestock Breeds and Organic Farming Systems. 2007. Available from: http://orgprints.org/10822/ [Accessed: 2015-04-11]

[79] Ángeles-Hernández JC, Perez-Hernandez AH, Pérez-Rocha MJ, González-Ronquillo M. Organic production of sheep milk. Tropical and Subtropical Agroecosystems. 2014;17(1):49–62.

[80] Tzouramani I, Sintori A, Liontakis A, Alexopoulos G. Assessing economic incentives for dairy sheep farmers: a real options approach. In: Proceeding of 12th Congress of

the European Association of Agricultural Economists (EAAE 2008); August 26–29; Ghent, Belgium; 2008.

[81] Kristensen ES, Thamsborg SM. Future European market for organic products from ruminants. In: Kyriazakis I, Zervas G, editors. Organic Meat and Milk from Ruminants. Netherlands: Wageningen Academic Publishers; 2002. p. 5–13. DOI: 10.3920/978-90-8686-506-2

[82] Christensen V, Saunders C. Economic analysis of issues concerning organic dairy farming. AERU Research Report No. 257, Lincoln, New Zealand: AERU, Lincoln University; 2003. Available from: https://researcharchive.lincoln.ac.nz/bitstream/handle/10182/746/aeru_rr_257.pdf?sequence=1. [Accessed: 2015-03-11]

[83] Dimitri C, Greene C. Recent growth patterns in the US organic foods market. Agriculture Information Bulletin. 2000, 777. Available from: http://www.ers.usda.gov/media/255736/aib777c_1_.pdf. [Accessed: 2015-03-18]

[84] Gerrard CL, Padel S, Moakes S. The use of farm business survey data to compare the environmental performance or organic and conventional farms. International Journal of Agricultural Management. 2012;2:5–16. DOI: 10.5836/ijam/2013-01-02

[85] Padel S, Lampkin N. Farm-level performance of organic farming systems. An overview. In: Lampkin N, Padel S, editors. The Economics of Organic Farming. Wallingford: CAB iInternational; 1994. p. 201–221.

[86] Zahm F, Viaux P, Girardin P, Vilain L, Mouchet C. Farm sustainability assessment using the IDEA method: from the 77 concept of farm sustainability to case studies on French farms. In: Proceedings and Outputs of the First Symposium of the International Forum on Assessing Sustainability in Agriculture (INFASA); 16 March; Bern, Switzerland; 2006.

[87] Hodges J. Session IV: role of international organizations and funding agencies in promoting gene-based technologies in developing countries. In: Makkar HPS, Viljoen GJ, editors. Applications of Gene-Based Technologies for Improving Animal Production and Health in Developing Countries. New York: Springer Science and Business Media, Inc.; 2005. p. 18–21 2005.

[88] Capper JL. The environmental sustainability of food production. In: Kebreab E, editor. Sustainable animal Agriculture. Wallingford, Oxfordshire: CABI International; 2014. p. 157–171. DOI: 10.1079/9781780640426.0000

4

Biochar Technology for Sustainable Organic Farming

Suarau O. Oshunsanya and OrevaOghene Aliku

Additional information is available at the end of the chapter

Abstract

The challenge of agricultural land depletion as a result of the pressure driven by the ever-growing population has brought about a renewed focus on the need for sustainable practices in agricultural production. Biochar is the solid carbonaceous product obtained when plant and/or animal biomass is subjected to pyrolysis. This chapter reviews the properties of biochar and its impacts when incorporated into the soil. Relative to its original organic form, this chapter iterates the benefits of biochar as a more sustainable organic approach towards improving agricultural soil qualities and hence crop yield due to its stability and duration in soils for hundreds of years. The impacts of biochar on soil physical, chemical and biological properties through the enhancement of soil nutrient and water-holding capacity, pH, bulk density and stimulation of soil microbial activities are by improving aggregation, porosity, surface area and habitat for soil microbes in biochar-amended soils. It is therefore recommended that biochar be used as soil amendment, especially to a degraded soil for a large and long-term carbon sink restoration.

Keywords: Biochar, Soil chemical properties, Soil water characteristics, Crop yield

1. Introduction

Throughout the world, intensive agriculture has often led to decline in soil physical, chemical and biological properties, leading to soil degradation. This decline in soil quality may be due to erosion and mining of nutrients and organic matter, hence preventing the soil from performing its functions such as regulating water flow, storing and cycling of nutrients, filtering, and transformation of organic and inorganic materials and sustaining biological productivity. However, considerably large amount of wastes such as crop residues, animal manure, etc. are being produced from many agricultural production systems. This organic waste may represent a considerable problem as well as new challenges and opportunities depending on how they are handled, which may determine whether there will be increase or decrease in biomass production, organic matter supply and decomposition rate.

In addressing the issue of decline in soil fertility, [1] reported that intentional and unintentional deposition of nutrient-rich materials on farmlands have in many cases led to an increase in soil fertility status. However, fresh residue materials have been reported to decompose until almost all carbon is lost [2]. This practice may not be sustainable when compared to the ever-growing human population per time. Thus, conversion of biomass to biochar could alter the transformation dynamics with respect to carbon sequestration. Soil carbon sequestration offers a large and long-term carbon sink to agricultural soils. Biochar is one of the sources of soil carbon sink, which could be obtained by subjecting biomass to pyrolysis. Pyrolysis is a process of combusting organic materials (biomass) under limited oxygen level [3].

Biochar as a soil amendment has become an important topic in soil science in the past few years, and the effects of biochar on agro-ecosystems are being studied by many researchers [4]. The conversion of biomass to bio-char as a carbon sink has been proposed before [5], but was not explicitly linked to an application to soil. As a soil amendment, biochar can greatly influence various soil properties and processes [6]. In fact, biochar may occur as a component of soil organic matter where slash-and-burn agriculture is widely practiced [7]. Many of the organic residues from agriculture, forestry and other production systems can be used to produce biochar and applied to agricultural soil both to sequester carbon and to improve the production potential of crops. This renewed focus in agriculture can be said to have started as a result of the discovery of the *Terra Preta de Indo* soils (Figure 1) located in the Amazon River Basin. From the assumptions surrounding the formation of the *Terra Preta* soils, agricultural scientists have come to believe that soil properties could be amended by applying biochar as an amendment [3]. Hence, biochar, the carbon-enriched, fine-grained product of biomass combusted under conditions of limited oxygen, is currently being widely studied for its effects as a soil amendment.

2. What is Biochar?

[8] defined biochar as a carbon-enriched, fine-grained and porous by-product of slow pyrolysis when organic material (feedstock) is thermally decomposed at low–moderate temperatures during long heating times under limited supply of oxygen. Feedstock may include wood materials, tree bark, crop residues, chicken litter, dairy manure or sewage sludge. Biochar is chemically and biologically more stable than the original fresh form from which it is produced due to its molecular configuration [9], making it more difficult to breakdown. This means that, in some cases, it can remain stable in soils for hundreds to thousands of years [10].

Source: [60]

Figure 1. Pictorial view of Latosol (left) and *Terra Preta* (right) soil horizon.

2.1. Properties of biochar

Biochars are characterized by certain morphological and chemical properties which are borne from the physico-chemical alteration of the original feedstock as a result of pyrolytic process. Characteristically, these properties of biochar differ since they are controlled by factors such as type of organic material from which they are made, pyrolysis conditions (i.e. final pyrolysis temperature or peak temperature, rate of heat application – slow or fast pyrolysis), rate and duration of charring [11,12,13]. The impact of biochar as an amendment depends on its properties. Key properties of biochar are the adsorptive properties that potentially alter soil's surface area, pore size distribution, bulk density, water-holding capacity and penetration resistance. Some physical properties of biochar determined by variations in feedstock type and pyrolysis condition are discussed below.

2.1.1. Large surface area and presence of micropores

Large surface area amendment property of biochar contributes to the adsorptive properties of soil and potentially improves pore size distribution, bulk density and consequently leading to an increase in the soil available water needed for crop growth and development. In addition, a strong direct relationship exists between a biochar's surface area and the pore volume as measured using N_2 adsorption and Braunauer-Emmett-Teller (BET) modelling [14,15]. [15] reported that the surface area could also be measured by using other compounds such as CO_2 on carbonaceous materials at the micrometer scale. [16] stated that understanding and determination of the relative abundance and stability of pores of different sizes are keys to soil ecosystem functioning. Important among these functions are aeration, hydrology and provision of habitat for microbes while the finer pores could be involved with molecular adsorption and transport [17].

Differences in production conditions, especially final combustion temperature, would result to variation in surface area of biochars even when they are produced from the same parent biomass. [16] stated that the relationship between the peak combustion temperature and surface morphological parameters (i.e. surface area, pore diameter and volume) of the resulting biochar is highly complex. [18] stated that there may be either no simple relationship between surface area and peak temperature, or surface area may increase with increase in peak temperature up to a certain threshold and then decrease. Due to variations in reports on surface area and peak temperature, [16] reported that the mechanisms responsible for increases in surface area with an increase in peak temperature or heating rate are not well understood. However, [11] reported that surface area increases with an increase in peak temperature of biochar production.

2.1.2. Adsorptive property

The adsorptive nature of biochar is related to its surface area. The adsorptive capability of biochar is determined by its surface chemical properties and porous nature. It is an important physical property due to its influence in the uptake and binding effect of materials from their surroundings [16]. [19] reported that biochar may adsorb poly aromatic compounds, poly aromatic and poly aliphatic hydrocarbons, other toxic chemicals, metals and elements or pollutants in soils, sediments, aerosols and water bodies.

2.1.3. Stability

This important physical property makes biochar a more sustainable soil amendment relative to its original fresh biomass for agricultural purpose. The evidence of high amounts of black carbon in the *Terra Preta* soils over a time suggests a high recalcitrant nature of biochar. However, degradation of at least some components (volatile matter or labile organic matter) of the biochar may occur [20]. On the other hand, [16] noted that the difference in sub-soil characteristics due to variations in microbial activity and oxygen content may affect biochar oxidation and aging. Biochar can move into sub-soil over time [21] to enrich the zone. Hence, other factors associated with its physical stability in soil include its mobility into deeper soil profile [16]. The aggregate stability of biochar-amended soil may also determine the susceptibility of biochars to microbial processes in subsoil. Mukherjee and Lal [16] explained that these factors not only enhance the stability of soil organic matter in the deeper profile but also improve availability of water and nutrients to crops and decrease erosion risks.

3. Restoring/improving soil properties

Biochar has the potential capacity to restore a degraded soil when added to the soil. Biochar mineralizes gradually over a long period of time when applied to the soil. Nutrients from biochar are released gradually to improve the physical, chemical and biological conditions of the soil. [12] reported that the impact of biochar as an amendment is a function of its properties such as large surface area and presence of micropores. These are key properties because they contribute to the adsorptive properties of soils and potentially alter soil physical and hydrological properties.

3.1. Biochar and soil properties

Figure 2 illustrates the interaction between biochar and soil. The application of biochar to the soil will alter the physical and chemical properties of the soil. [22] stated that the net effect of biochar on the soil physical properties will depend on its interaction the physico-chemical characteristics of the soil, the weather conditions prevalent at the particular site and the management of its application. Biochar application can reduce the bulk density of the different soils [23]. This could bring about improvement in soil structure or aggregation, and aeration enhancement, thus improving soil porosity. [17] reported that the higher the total porosity (micro- and macropores) the higher is soil physical quality. This is because micropores are involved in molecular adsorption and transport of water and nutrients while macropores affect aeration and drainage. Several studies have reported that as low as 0.5% (g g^{-1}) biochar application rate was sufficient to improve water-holding capacity and water retention [24,25]. Hence, this can be said to be good water-holding capacity amendment for sandy soils which are highly porous due to the preponderance of macropores.

3.2. Effect of biochar application on some soil physical properties

A key determinant of soil functions and processes is its physical properties, precisely and most importantly, its texture. Hence, the addition of biochar in soils with different textures should affect the soil hydraulic properties differently due to the fact that there is a correlation between soil texture and soil hydraulic properties. The impacts of biochar as a soil amendment on some soil physical and hydrological properties are briefly discussed below.

3.2.1. Soil surface area

Table 1 depicts a summary of results of biochar application on surface area. Soil surface area is an intrinsic property of soil determined by the sizes of its particles. The surface area of soils

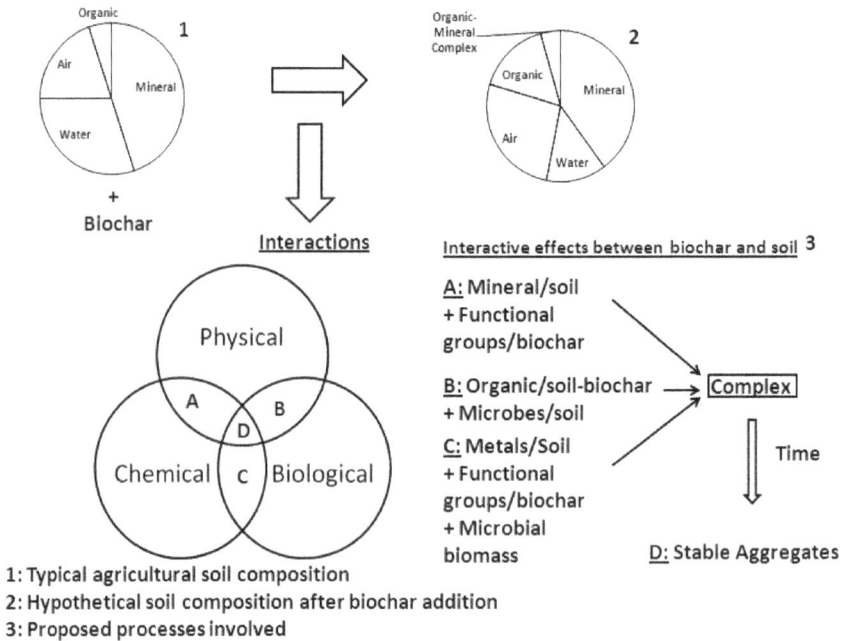

Figure 2. Schematic representation of interactions between biochar and soil [16].

is an important physical characteristic which plays a vital role in water- and nutrient-holding capacities, aeration and microbial activities [26]; hence, it can be said to be partly controlling the essential functions of soil fertility. However, the effectiveness of the surface area of a soil depends on its size – the larger the surface area, the greater the soil's water- and nutrient-holding capacities. This is particularly true for fine-textured soils. Thus, [16] reported that agronomic productivity improvement of biochar-amended soils may be linked to the higher surface area of the biochar–soil mixtures. [17,27,28] explained that the high surface area of biochar provides the space for formation of bonds and complexes with cations and anions with metals and elements of soil on its surface, which improves the nutrient retention capacity of soil. [28] reported that biochar incorporation can enhance specific surface area up to 4.8 times that of adjacent soils. [29] also reported increases in specific surface area of an amended clayey soil from 130 to 150 m^2 g^{-1} when biochar derived from mixed hardwoods was applied at rates of 0 to 20 g kg^{-1} in a long-term soil column incubation study.

3.2.2. Porosity

Table 1 shows a summary of results of biochar application on soil porosity. This is the ratio of the pore volume to the total volume of a representative sample of a porous medium. This factor is said to be associated with surface area. The total porosity or pore size distribution of biochar is a factor that can play an important role in the alteration of the properties of biochar-amended soils. Biochars are usually characterized by the preponderance of micropores, which may alter the pore size distribution of coarse texture soil when added. [24] reported that significant increases in mesoporosity occurred at the expense of macropores in waste-derived biochar-amended soil compared to the control. [24] further intensified that the higher the rate of biochar application the greater its effect on porosity. Hence, biochar could be a good replacement for tillage practices which causes short-term increase in porosity, but long-term decrease in aggregation and ultimately lowering soil porosity.

Soil type	Biochar type	Study type (scale)	Rate of biochar application	SA	Porosity	Reference
			$(g\,g^{-1})$	$(m^2\,g^{-1})$	(%)	
Residue sand	Municipal green waste, 450°C	Laboratory	0	-	0.46	[24]
			2.6	-	0.48	
			5.2	-	0.51	
Clarion fine loamy	Mixed hardwoods (Oak: Quercus spp., Hickory: Carya spp.), 500°C	Laboratory	0	130	-	[29]
			0.5	133	-	
			1.0	138	-	
			2.0	153	-	
Sandy soil	Jarrah woods (Eucalyptus marginata), 600°C	Greenhouse	0	1.3	56.1	[61]
			0.45	2.7	57.6	
			2.27	8.4	62.1	
Silt loam	Birch (Betula pendula), 400°C	Field	0	-	50.9	[62]
			1.2	-	52.8	

Table 1. Impact of biochar on Surface area (SA) and porosity of amended soils

3.2.3. Bulk density

Table 2 shows the results of biochar application on soil bulk density. Bulk density, which is defined as the mass of soil per its unit volume, has been known to have a negative correlation with surface area. [30] stated that well-structured soils (fine texture) are characterized by low bulk density values between 1.0 and 1.3 g cm^{-3} while poorly structured (coarse texture) soils are known to have high bulk density values between 1.6 and 1.8 g cm^{-3}. Hence, reports from both field and laboratory studies have shown bulk densities to have contrasting results to surface areas of biochar-amended soils. [29], [24] and [23] reported that application of biochar can decrease the bulk density of soils. [29] showed in a soil column incubation study that biochar-amended soil columns had significantly lower bulk density than no-biochar controls. [16] reported that biochar-amended column had a lower rate of compaction compared to the control or manure-amended soil columns when all the columns were subjected to compaction by gravity and periodical leaching events. They further stated that the decrease in bulk density of biochar-amended soil could be one of the indicators of the improvement of soil structure or aggregation and aeration, and could be soil-specific.

Soil types	Biochar type	Study type (scale)	Rate of biochar application	Bulk density	Reference
			% (g g^{-1})	g cm^{-3}	
Norfolk loamy sand: E	Pecan (*Carya illinoinensis*) shells, 700°C	Laboratory	0	1.52	[32]
			2.1	1.45^1, 1.52^2	
Norfolk loamy sand: E and Bt			0	1.34	
			2.1	1.36^1, 1.34^2	
Hydroagric stagnic anthrosol	Wheat (*Triticum* spp.) straw, 350–550°C	Field	0	0.99, 0.94^3	[63]
			1.1	0.96, 0.91^3	
			2.2	0.91, 0.86^3	
			4.4	0.89, 0.88^3	
Residue sand	Municipal green waste , 450°C	Laboratory	0	1.65	[24]
			2.6	1.55	
			5.2	1.44	
Clarion fine loamy	Mixed hardwoods, 500°C	Laboratory	0	1.21, 1.34^4	[29]
			0.5	1.10, 1.24^4	
			1.0	1.08, 1.24^4	
			2.0	1.08, 1.24^4	

Source: [16]. [1] measured after 44 days; [2] measured after 94 days; [3] measured after 1 year; [4] measured after 15 months.

Table 2. Soil bulk density as affected by biochar application

3.2.4. Aggregate stability

Results of studies showing biochar effect on soil aggregation are illustrated in Table 3. Studies have shown biochar to respond positively to aggregation. Though [16] reported that data on aggregate stability and penetration resistance of biochar-amended soils are scarce, a few studies generally showed that low-temperature (220°C) hydrochar made from spent brewer's grains (a residue from beer brewing) responded positively to aggregation of Albic Luvisol by significantly increasing water-stable aggregates as compared to the control treatment. [31] have reported that the formation of complexes of biochar with minerals, as the result of interactions between oxidized carboxylic acid groups at the surface of biochar particles, should be responsible for the improved soil aggregate stability (Figure 2). As a result, soil aggregates and pore size distribution can be improved by adding organic matter from biodegradation and thus improving soil hydraulic properties. However, other authors have reported contrasting results. For instance, [32] reported that with or without mixing Bt and E horizons with pecan shell (*Carya illinoinensis*), biochar-amended soil decreased aggregation compared to the control, while [33] reported mixing of biochar from pecan with switchgrass increased aggregation, but the effect was however significantly lower when the soil was treated only with biochar without mixing with switchgrass. From this trend of results, [16] concluded that a positive effect on soil aggregate stability would require the presence of a substrate (i.e switchgrass) along with biochar as an amendment.

3.2.5. Penetration resistance

Studies on the effect of biochar amendment on soil penetration resistance are illustrated in Table 3. Penetration resistance measures the capacity of a soil in its confined state to resist penetration by a rigid object [34]. It is affected by moisture content. Thus, it affects the potential for root growth and development. Ehlers et al. [35] found root growth to be inversely related to penetration resistance. Results from literatures have shown that the effect of biochar application on soil penetration resistance is dependent on time of application. Busscher et al. [32] reported that mixing Norfolk loamy sand E and E and Bt layers with pecan shell biochar produced at a temperature of 700°C increased penetration resistance measured after 44 days of application. Penetration resistance was, however, reduced when measured after 96 days of application. Thus, soil compaction may not be alleviated by biochar addition over short period of time, but may be altered in the long run due to changes in properties as a result of aging of biochar.

Soil types	Biochar type	Study type (Scale)	Rate of biochar application	Aggregation	Penetration resistance	Reference
			% (g g^{-1})	%	MPa	
Norfolk loamy sand: E	Pecan shells, 700°C	Laboratory	0	14.3	1.19[1], 0.80[2]	[32]
			2.1	12.9	1.27[1], 0.88[2]	
Norfolk loamy sand: E and Bt			0	27.3	0.71[1], 0.76[2]	
			2.1	20.9	0.88[1], 0.94[2]	
Norfolk loamy sand: Ap	Pecan shells, 700°C	Laboratory	0	9.95, 13.0*	1.04[1], 1.1[2]	[33]
			0.5	9.53, 12.7*	0.96[1], 1.15[2]	
			1.0	10.7, 12.3*	1.03[1], 1.02[2]	

Soil types	Biochar type	Study type (Scale)	Rate of biochar application	Aggregation	Penetration resistance	Reference
			2.0	9.23, 11.8[*]	0.82[1], 0.87[2]	
Albic Luvisol	Hydrochar, 220°C	Laboratory	0	49.8	-	[64]
			5	69.0	-	
			10	65.1	-	
		Greenhouse	0	10.3	-	
			5	20.8	-	
			10	33.8	-	

Table 3. Soil aggregation and penetration resistance as affected by biochar application

3.3. Hydrological properties

Several authors have reported positive response of soil hydrological properties to biochar amendment. This may be due to the fact that soil hydrological properties such as infiltration rate, moisture content, hydraulic conductivity, water-holding capacity and water retention are invariably related to soil surface area, bulk density, porosity and aggregate stability [16]. In other words, an alteration in these soil physical properties as caused by biochar application would lead to a change in soil hydrological properties.

3.3.1. Water-holding capacity, water retention and moisture content

Table 4 shows the results of biochar application effect on water-holding capacity. The amount of water in a soil is a function of its ability to hold and retain water for plant use against the influence of gravity. Fine-textured soils would have higher moisture content at the same tension as soils with coarse particles. This is because the ability of a soil to retain water is a function of the micropores in the soil, which is usually lower in coarse-textured soils. Hence, moisture required by plants to upset the evapotranspirational demand of the atmosphere may be limiting, especially in coarse-textured soils. Thus, application of biochar can increase water-storage ability of coarse-textured soils. Several studies have reported alterations in water-holding capacity and water retention in soils amended with biochar. [33] and [36] reported that 0.5% (g g^{-1}) biochar application rate was sufficient to improve water-holding capacity. Application of biochar produced from black locust (*Robinia pseudoacacia*) was reported to increase the available water capacity by 97%, saturation water content by 56%, but reduced hydraulic conductivity [25]. This can also influence soil aeration and temperature to a very large extent. [29] reported that results from a long-term column study indicated that biochar-amended Clarion soil retained up to 15% more water, with 13% and 10% more water retention at −100 KPa and −500 KPa soil matric potential, respectively, compared to control (unamended soils). [37] showed that coal-derived humic acid substances can increase water retention, available water capacity and aggregate stability of inherently degraded soils. [38] reported that biochar application increased the available water capacity in sandy soil, with no effect on a loamy soil, and decreased moisture content in a clayey soil. [16] suggested that such response may be due to the hydrophobic nature of the charcoal that caused alterations in soil pore size distribution. [38], therefore, advised that because the soil moisture retention may only be improved in coarse-textured soils, a careful choice of biochar/soil combination needs to be taken into consideration.

Soil types	Biochar type	Study type (Scale)	Rate of biochar application	Water holding capacity	Reference
			% (g g^{-1})	(g cm^{-3})	
Residue sand	Municipal green waste, 450°C	Laboratory	0	0.11	[24]
			2.6	0.16	
			5.2	0.20	
Norfolk loamy sand: Ap	Pecan shells, 700°C	Laboratory	0	0.64	[33]
			0.5	0.59	
			1.0	0.60	
			2.0	0.66	
Sandy loam	Ponderosa pine (*Pinus ponderosa*), 450°C	Laboratory	0	11.9	[36]
			0.5	12.4	
			1.0	13.0	
			5.0	18.8	
Silt loam	Birch, 400°C	Field	0	0.49	[62]
			1.2	0.54	

Table 4. Soil water holding capacity as affected by biochar application

3.4. Biochar and soil chemical properties

Most studies of biochar as a soil amendment have focused majorly on soil nutrient status, taking into consideration cation exchange capacity, nutrient content, pH, the carbon sequestration potential of the amended soil, and vegetative growth and yield of crops. Biochar has the potential to improve soil CEC due to the fact that it is often characterized by high CEC values, due to its negative surface charges and its high specific surface area as was reported for biochar produced from crop residues [39].

Furthermore, the immediate beneficial effect of biochar application on crop productivity in tropical soils may result from increase in availability of nitrogen, phosphorus, potassium, calcium, copper and zinc as reported for soils amended with secondary forest biochar [40]. Also, poultry litter biochar may result in strong increase in soil extractable phosphorus [41] when incorporated into the soil. In evaluating the effect of different biochars on soil chemical properties, [42] reported that biochar produced from poultry manure had higher electrical conductivity, nitrogen, phosphorus and pH values than that of garden waste. However, this may be due to their effects in reducing leaching and fixation of nutrients as moderate biochar additions are not a direct supplier of plant nutrients in the long-term.

3.5. Effects of biochar application on Soil Organic Carbon (SOC)

Biochar application can directly or indirectly affect SOC dynamics. Indirectly, biochar could affect net primary production and, thus, the amount of biomass that may remain in agro-ecosystems. This would result to alteration in soil carbon inputs. [8] stated that higher below-ground net primary production and increased root-derived carbon inputs after biochar application may particularly result in an increase in SOC.

Directly, biochar can inhibit degradation process, and as a result increase the mean residence time (MRT) of SOC (i.e. the mean time that a SOC-carbon atom spends in soil). As a direct consequence, biochar application would enhance SOC stabilization processes and contribute to SOC sequestration. The MRT of biochar-carbon is thought by some to be in the range of millennia [43]. However, information on biochar longevity in soil is meagre and varies between biochars and sites. For example, the MRTs of biochar in field experiments ranged from about 8 years for biochar produced by burning of forest trees during slash-and-burn agricultural practices [44] to 3,600 years for biochar produced from prunings of old mango (*Mangifera indica* L.) trees [45]. Also, biochar longevity in soil may be affected by differences in climatic conditions. For example, chemical and/or biological mineralization of natural chars produced from wood during bushfires was slower under Mediterranean climate when compared to temperate climates in Australia [46].

3.6. Liming effect

Biochar can be said to be acidic or alkaline in nature depending on the temperature of the materials used during pyrolysis. [47] explained that the acid functional group concentration in biochars produced from the biomass of rice, valley oak (*Quercus lobata Ne´ e*), etc decreased with increasing peak pyrolysis temperature as more fused aromatic ring structures were produced and more volatile matter was lost. The effectiveness of both types will depend on the pH of the soil to be amended. [48] stated that the alkaline biochars produced at higher pyrolysis temperature are more effective in supporting increases in biomass by improved growth conditions than acidic biochars presumably through increases in soil alkalinity. [49] stated that the moderation in aluminium toxicity may be the reason why biochar application has positive effects on productivity in tropical and irrigated systems on highly weathered and acid soils with low-activity clays. This is because the reduction of aluminium and iron concentrations in the soil solution will enhance the availability of previously bound phosphorus to plants, and plant roots would be able to explore even acid soils to absorb nutrients and water more effectively.

3.7. Effect of biochar on soil microorganisms

Studies have shown higher microbial biomass but yet lower microbial activity in biochar-amended soil than the neighbouring soils [50]. However, most studies have focused on biochar interaction with mycorrhizal fungi [50]. Specifically, biochar has been reported to have symbiotic relationship with the mycorrhizal system. According to [51], the four mechanisms by which biochar could improve mycorrhizal abundance (40%) and functioning are listed as follows:

i.   Alteration of soil physico-chemical properties,

ii.   Indirect effects on mycorrhizae through effects on other soil microbes,

iii.   Plant-fungus signalling interference, and

iv.   Detoxification of allelochemicals on biochar.

[52] noted 50% to 72% increase in soil biological nitrogen fixation through biochar application. [53] have hypothesized both bacteria and fungi to be better protected from grazers or competitors by exploring pore habitats in biochars. This is because biochar provides microbial habitat and refugia for microbes where they are also protected from unfavourable conditions.

3.8. Effect of biochar on crop yield

The summary of experiments assessing the impact of biochar addition on crop yield is showed in Table 5. From the agricultural perspective, the summary of the effect of biochar in regulating soil hydrological, physical and chemical properties results to improved soil productivity and consequently increased crop yield. However, the effect of biochar on soil health as well as crop productivity can be influenced by the forms (dust, fine particles, coarse grain) and the methods of application (surface application, top dressing, drilling) of biochar to soil. [54] clearly explained that even small quantities of biochar added to seed coatings may in some cases be sufficient for a beneficial effect.

[40] reported increasing crop yields with increasing biochar applications of up to 140 t carbon ha^{-1} on highly weathered soils in the humid tropics. Also, [55] found that the biomass growth of beans rose with biochar applications up to 60 t carbon ha^{-1}. Furthermore, scientists have reported that application of biochar on soil has significant effect on net primary crop production, grain yield and dry matter production [56,57,58,59].

Author	Study outline	Results summary
[65]	Cowpea on xanthic ferralsol	Char at 67 t/ha increased biomass by 150%
		Char at 135 t/ha increased biomass by 200%

Author	Study outline	Results summary
[40]	Soil fertility and nutrient retention. Cowpea was planted in pots and rice crops in lysimeters, Brazil	Biochar additions significantly increased biomass production by 38% to 45% (no yield reported)
[66]	Comparison of maize yields between disused charcoal production sites and adjacent fields, Ghana	Grain and biomass yield was 91% and 44% higher on charcoal site than control
[67]	Maize, cowpea and peanut trial in area of low soil fertility	*Acacia* bark charcoal plus fertilizer increased maize and peanut yields (but not cowpea)
[42]	Pot trial on radish yield in heavy soil using commercial green waste biochar (three rates) with and without nitrogen	Biochar at 100 t/ha increased yield ×3; linear increase 10 to 50 t/ha, but no effect without added nitrogen

Source: [16].

Table 5. Summary of experiments assessing the impact of biochar addition on crop yield

4. Conclusion

Biochar, as an amendment on soil physical, chemical and biological properties, depends on environmental conditions, dynamic properties of soils, biochar properties which are a function of the organic materials and conditions used for biochar production and the rate and method of application.

Notable soil physical properties found to be enhanced by biochar include soil surface area, bulk density, porosity, aggregate stability, penetration resistance and moisture content. Also, soil pH, organic carbon and cation exchange capacity were enhanced in biochar-amended soils. Biologically, mycorrhizal abundance, biological nitrogen fixation, microbial biomass and microbial habitats were improved in biochar-amended soils compared to unamended soils.

Modification of soil physical, chemical and biological properties by biochar application resulted to improved plant nutrient retention, acquisition and availability, leading to improved biomass growth, dry matter production and crop yields.

Author details

Suarau O. Oshunsanya and OrevaOghene Aliku

*Address all correspondence to: soshunsanya@yahoo.com

Department of Agronomy, University of Ibadan, Ibadan, Nigeria

References

[1] Woods WI. Slash and Char. In: Balance Carbon and Restore Soil Fertility [Internet]. 2003 [Updated: 2006]. Available from: www.biochar.org [Accessed: 2 March, 2015]

[2] Lehmann J, Gaunt J, Rondon M. Bio-char sequestration in terrestrial ecosystems – a review. Mitig Adapt Strategies Glob Change. 2006;11:395–419.

[3] Fagbenro JA, Oshunsanya SO, Oyeleye BA. Effects of gliricidia biochar and inorganic fertilizer on moringa plant grown in an oxisol. Commun Soil Sci Plant Analysis 2015;46(5):619–26.

[4] Anders E, Watzinger A, Rempt F, Kitzler B, Wimmer B, Zehetner F, et al. Biochar affects the structure rather than the total biomass of microbial communities in temperate soils. Agri Food Sci 2013;22:404–23.

[5] Seifritz W. Should we store carbon in charcoal? Int J Hydrogen Energy 1993;18:405–7.

[6] Lehmann J, Joseph S. Biochar for Environmental Management: Science and Technology. London: Earthscan; 2009.

[7] Glaser B, Haumaier L, Guggenberger G, Zech W. The 'terra preta' phenomenon: a model for sustainable agriculture in the humid tropics. Naturwissenschaften 2001;88:37–41.

[8] Sohi S, Krull E, Lopez-Capel E, Bol R. A review of biochar and its use and function in soil. Adv Agron 2010;105:47–82.

[9] Krishnakumar S, Kumar SR, Mariappan N, Surendar KK. Biochar-boon to soil health and crop production. Afr J Agri Res 2013;8(38):4726–39.

[10] Krull ES, Swanston CW, Skjemstad JO, McGowan JA. Importance of charcoal in determining the age and chemistry of organic carbon in surface soils. J Geophys Res. 2006;P. 111 G04001DOI: doi:10.1029/2006JG000194

[11] Mukherjee A. Physical and Chemical Properties of a Range of Laboratory-produced Fresh and Aged Biochars [thesis]. Gainesville, FL, USA: University of Florida; 2011.

[12] Mukherjee A, Zimmerman AR, Harris WG. Surface chemistry variations among a series of laboratory-produced biochars. Geoderma 2011;163:247–55.

[13] Zimmerman AR. Abiotic and microbial oxidation of laboratory-produced black carbon (biochar). Environ Sci Technol 2010;44:1295–301.

[14] Sweatman MB, Quirke N. Characterization of porous materials by gas adsorption: Comparison of nitrogen at 77 k and carbon dioxide at 298 k for activated carbon. Langmuir 2001;17:5011–20.

[15] Jagiello J, Thommes M. Comparison of DFT characterization methods based on N2, Ar, CO2, and H2 adsorption applied to carbons with various pore size distributions. Carbon 2004;42:1227–32.

[16] Mukherjee A, Lal R. Biochar impacts on soil physical properties and greenhouse gas emissions. Agronomy 2013;3:313–39.

[17] Atkinson CJ, Fitzgerald JD, Hipps NA. Potential mechanisms for achieving agricultural benefits from biochar application to temperate soils: a review. Plant Soil 2010;337:1–18.

[18] Fernandes MB, Skjemstad JO, Johnson BB, Wells JD, Brooks P. Characterization of carbonaceous combustion residues. In: Morphological, elemental and spectroscopic features. Chemosphere 2003;51:785–95.

[19] Schmidt MWI, Noack AG. Black carbon in soils and sediments: analysis, distribution, implications, and current challenges. Glob Biogeochem Cycles. 2000;14:777–93.

[20] Hammes K, Torn MS, Lapenas AG, Schmidt MWI. Centennial black carbon turnover observed in a Russian steppe soil. Biogeosciences 2008;5:1339–50.

[21] Skjemstad JO, Taylor JA, Janik LJ, Marvanek SP. Soil organic carbon dynamics under long-term sugarcane monoculture. Aust J Soil Res 1999;37:151–64.

[22] Verheijen F, Jeffery S, Bastos AC, Van der Velde M, Diafas I. Biochar Application to Soils: A Critical Scientific Review of Effects on Soil Properties, Processes and Functions. JRC Scient. Technical reports. 2010; 1-166.

[23] Chen H-X, Du Z-L, Guo W, Zhang Q-Z. Effects of biochar amendment on cropland soil bulk density, cation exchange capacity, and particulate organic matter content in the North China plain. Yingyong Shengtai Xuebao. 2011;22:2930–4.

[24] Jones BEH, Haynes RJ, Phillips IR. Effect of amendment of bauxite processing sand with organic materials on its chemical, physical and microbial properties. J Environ Manag 2010;91:2281–8.

[25] Uzoma KC, Inoue M, Andry H, Zahoor A, Nishihara E. Influence of biochar application on sandy soil hydraulic properties and nutrient retention. J Food Agric Environ 2011;9:1137–43.

[26] Van Zwieten L, Singh B, Joseph S, Kimber S, Cowie A, Yin Chan K. Biochar and emissions of non-CO_2 greenhouse gases from soil. In: Lehmann J, Joseph S. (eds.) Biochar for Environmental Management. VA, USA: Earthscan: London Sterling; 2009.

[27] Hammes K, Schmidt, M. Changes in biochar in soil. In: Lehmann J, Joseph S. (eds.) Biochar for Environmental Management: Science and Technology. London, UK: Earthscan: 2009. pp. 169–182.

[28] Liang B, Lehmann J, Solomon D, Kinyangi J, Grossman J, O'Neill B, et al. Black carbon increases cation exchange capacity in soils. Soil Sci Soc Am J 2006;70:1719–30.

[29] Laird DA, Fleming P, Davis DD, Horton R, Wang BQ, Karlen DL. Impact of biochar amendments on the quality of a typical midwestern agricultural soil. Geoderma 2010;158:443–9.

[30] Oshunsanya SO. Soil Physics. 1st ed. Nigeria: Dabank Publishers; 2011. 166 p.

[31] Glaser B, Lehmann J, Zech W. Ameliorating physical and chemical properties of highly weathered soils in the tropics with charcoal - a review. Biol Fert Soils 2002;35:219–30.

[32] Busscher WJ, Novak JM, Ahmedna M. Physical effects of organic matter amendment of a southeastern us coastal loamy sand. Soil Sci 2011;176:661–7.

[33] Busscher WJ, Novak JM, Evans DE, Watts DW, Niandou MAS, Ahmedna M. Influence of pecan biochar on physical properties of a norfolk loamy sand. Soil Sci 2010;175:10–4.

[34] Soil Survey Division Staff. Soil Survey Manual. Washington, D.C.: USDA-NRCS; 1993.

[35] Ehlers W, Popke V, Hesse F, Bohm W. Penetration resistance and root growth of oats in tilled and untilled loam soil. Soil Tillage Res 1983;3:261–75.

[36] Briggs C, Breiner JM, Graham RC. Physical and chemical properties of pinus ponderosa charcoal: implications for soil modification. Soil Sci 2012;177:263–8.

[37] Minamikawa K, Hayakawa A, Nishimura S, Akiyama H, Yagi K. Comparison of indirect nitrous oxide emission through lysimeter drainage between an andosol upland field and a fluvisol paddy field. Soil Sci Plant Nutr 2011;57:843–54.

[38] Tryon EH. Effect of charcoal on certain physical, chemical, and biological properties of forest soils. Ecol Monogr 1948;18:81–115.

[39] Yuan JH, Xu RK, Zhang H. The forms of alkalis in the biochar produced from crop residues at different temperatures. Biores Technol 2011;102:3488–97.

[40] Lehmann J, da Silva JP, Steiner C, Nehls T, Zech W, Glaser B. Nutrient availability and leaching in an archaeological anthrosol and a ferralsol of the Central Amazon basin: fertilizer, manure and charcoal amendments. Plant Soil 2003;249:343–57.

[41] Novak JM, Busscher WJ, Laird DA, Ahmedna M, Watts DW, Niandou MAS. Impact of biochar amendment on fertility of a Southeastern coastal plain soil. Soil Sci 2009;174:105–12.

[42] Chan KY, Van Zwieten L, Meszaros I, Downie A, Joseph S. Assessing the agronomic values of contrasting char materials on an Australian hard setting soil. In: International Agrichar Initiative (IAI); 27 April–2 May 2007; Terrigal, New South Wales. Australia: 2007.

[43] Glaser B, Birk JJ. State of the scientific knowledge on properties and genesis of Anthro-pogenic Dark Earths in Central Amazonia (*terra preta de I´ndio*). Geochim Cosmochim Ac 2013;82:39–51.

[44] Nguyen BT, Lehmann J, Kinyangi J, Smernik R, Riha SJ, Engelhard MH. Long-term black carbon dynamics in cultivated soil. Biogeochemistry 2008;89:295–308.

[45] Major J, Rondon M, Molina D, Riha SJ, Lehmann J. Maize yield and nutrition after 4 years of doing biochar application to a Colombian savanna oxisol. Plant Soil 2010;333:117–28.

[46] McBeath AV, Smernik RJ, Krull ES. A demonstration of the high variability of chars produced from wood in bushfires. Org Geochem 2013;55:38–44.

[47] Li X, Shen Q, Zhang D, Mei X, Ran W, Xu Y, et al. Functional groups determine biochar properties (pH and EC) as studied by two-dimensional 13C NMR correlation spectrosco-py. PLoS ONE 2013;8DOI: DOI: 10.1371/journal.pone.0065949

[48] Biederman LA, Harpole WS. Biochar and its effects on plant productivity and nutrient cycling: a meta-analysis. GCB Bioenerg 2013;5:202–14.

[49] Blackwell P, Riethmuller G, Collins M. Biochar application to soil (Chapter 12). In: Lehmann J, Joseph S. (eds.) Biochar for Environmental Management: Science and Technology. London, UK: Earthscan; 2009. p. 207.

[50] Thies JE, Rillig M. Characteristics of biochar: biological properties. In: Lehmann J, Joseph S. (eds.) Biochar for Environmental Management: Science and Technology. London: Earthscan; 2009. pp. 85–105.

[51] Warnock DD, Lehmann J, Kuyper TW, Rillig MC. Mycorrhizal responses to biochar in soil—concepts and mechanisms. Plant Soil 2007;300:9–20.

[52] Lehmann J, Rondon M. Biochar soil management on highly weathered soils in the humid tropics. In: Uphoff N, et al. (eds.) Biological Approaches to Sustainable Soil Systems. Boca Raton, FL: CRC Press; 2006. pp. 517–30.

[53] Saito M, Marumoto, T. Inoculation with arbuscular mycorrhizal fungi: the status quo in Japan and the future prospects. Plant Soil 2002;244:273–9.

[54] Hill RA, Harris A, Stewart A, Bolstridge N, McLean KL, Blakeley R. Charcoal and se-lected beneficial microorganisms: plant trials and SEM observations. In: International Agrichar Conference; 2007; Terrigal May. NSW Australia: 2007.

[55] Rondon M, Lehmann J, Ramirez J, Hurtado M. Biological nitrogen fixation by com-mon beans (Phaseolus vulgaris L.) increases with biochar additions. Biol Fertil Soils 2007;43:688–708.

[56] Chan Y, Van Zwieten L, Meszaros I, Downie A, Joseph S. Using poultry litter bio-chars as soil amendments. Austr J Soil Res 2008;46:437–44.

[57] Chan KY, Xu Z. Biochar: nutrient properties and their enhancement. In: Lehmann J, Joseph S. (eds.) Biochar for environmental management. Science and Technology, London: Earthscan; 2009. pp. 67–84.

[58] Major J, Steiner C, Downie A, Lehmann J. Biochar effects on nutrient leaching. In: Lehmann J, Joseph S. (eds.) Biochar for environmental management. Science and Technology. London: Earthscan ; 2009. pp. 271–287.

[59] Spokas KA, Reicosky DC. Impacts of sixteen different biochars on soil greenhouse gas production. Ann Environ Sci 2009;3:179–93.

[60] Anderson PS, McLaughlin H. Biochar basics: an introduction about the "what and why" of biochar. In: Northeast Biochar Symposium; 2009; USA. University of Massachussetts Amherst: 2009.

[61] Dempster DN, Gleeson DB, Solaiman ZM, Jones DL, Murphy DV. Decreased soil microbial biomass and nitrogen mineralisation with eucalyptus biochar addition to a coarse textured soil. Plant Soil 2012;354:311–24.

[62] Karhu K, Mattila T, Bergstro°m I, Regina K. Biochar addition to agricultural soil increased CH_4 uptake and water holding capacity — results from a short-term pilot field study. Agric Ecosyst Environ 2011;140:309–13.

[63] Mankasingh U, Choi PC, Ragnarsdottir V. Biochar application in a tropical, agricultural region: a plot scale study in Tamil Nadu, India. Appl Geochem 2011;26:S218–21.

[64] George C, Wagner M, Kucke M, Rillig MC. Divergent consequences of hydrochar in the plant-soil system: Arbuscular mycorrhiza, nodulation, plant growth and soil aggregation effects. Appl Soil Ecol 2012;59:68–72.

[65] Glaser B, Lehmann J, Zech W. Ameliorating physical and chemical properties of highly weathered soils in the tropics with charcoal – a review. Biol Fertil Soils 2002;35:219–30.

[66] Oguntunde PG, Fosu M, Ajayi AE, Van de Giesen N. Effects of charcoal production on maize yield, chemical properties and texture of soil. Biol Fertil Soils 2004;39:295–9.

[67] Yamamoto M, Okimori Y, Wibowo IF, Anshori S, Ogawa M. Effects of the application of charred bark of *Acacia mangium* on the yield of maize, cowpea and peanut, and soil chemical properties in South Sumatra, Indonesia. Soil Sci Plant Nutr 2006;52:489–95.

5

Abundance and Risk Factors for Dermatobiosis in Dairy Cattle of an Organic Farm in the Tropical Region

Mônica Mateus Florião and Wagner Tassinari

Additional information is available at the end of the chapter

Abstract

Studies about *Dermatobia hominis* larvae have been described, but no data were found regarding dairy cattle from organic production system in tropical region. The herd consisted of 40 dairy crossbred zebu x taurine. Fortnightly inspection (915 inspections) with mapping for the presence of larvae in the body surface was carried out over the period of a year in the area of the Integrated Agroecological Production System –IAPS/RJ – a technical cooperation project. The results indicated predominance of parasitism in females (average 21.98). In males, the highest number of nodules were on the right side (4.46); in females, highest number of nodules were on the left side. The infestation in adults (average 31.55) was highest; animals in lactation were less infested (average 8.01); in young animals, the most infested side was the left; the most infested coat was the black on white (average 36.69); the less infested coats were red with typical shades (average 14.13) and light brown and dark (12.33). Each increment of 1 mm³ of water caused a mean increase of 1.03 in the relative risk of occurrence of dermatobiosis and with every one degree increased there was an average increase of 1.14 in the relative risk for infestation.

Keywords: Nodular subcutaneous myiasis, organic management, bovine

1. Introduction

Dermatobia hominis (Linneus Jr., 1781) (Diptera: Cuterebridae), commonly known in Brazil as "mosca do berne" (warble fly), has high incidence in cattle bred in many regions of the country. It infests a considerably large number of hosts, cattle being the most affected. This fly's larva once on the skin of those animals causes furuncular myiasis, also known as dermatobiosis, which is characterized by the formation of nodules in the host.

The study of seasonal variations of this fly allows us to know the periods of higher parasitic intensity, and also to correlate the facts operating in the growth of its population. Several authors are engaged in studying the seasonal variation of fly's larva and its relation to climate elements: temperature, precipitation, and humidity, showing that the presence of the warble is associated with regions that have moderately high temperatures during the day and relatively cold overnights, median and abundant rainfall, dense vegetation, and a considerable number of animals. Also, the rainy season is the period of highest occurrence. Even with all these characteristics, the index of parasitism by *D. hominis* can vary according to environmental conditions, regional differences. It also depends on the location of the parasite in the host.

In conventional livestock, the larva population on the cattle is controlled with the use of chemical larvicide. On the other hand, organic rural properties must meet the standards contained in the 60th Article of Normative Instruction No. 46, 2011, Ministry of Agriculture and Supply, which regulates organic production in Brazil, restricting the use of allopathic medicines [1].

Several studies about seasonality and *D. hominis* larvae's control have already been described, but no data were found related to this infestation in dairy cattle raised under organic systems. The goal of this study was to provide subsides about this parasitic skin disease in organic breeding; to verify the location and distribution of the larvae on the body surface of the cattle; to determine the intensity of infestation related to gender, age, and coat color; and the influence of the climatic factors in infestation rates. This study also provides basis to the creation of a dermatobiosis control program in organic dairy production systems.

2. Literature review

The parasitism rate of *D. hominis* may have some variations due to climate conditions. In addition, there may be differences in the location of the parasite in the host.

2.1. *Dermatobia hominis*: Geographical distribution and biology

According to [2], flies of the species *D. hominis* are diurnal and are found in tropical forests. According to [3], flies of *D. hominis* were never found in stables and houses, being more abundant on the edges of woods, forests, and eucalyptus plantations. As [4] says, this fly is well adapted in Brazil, mainly concentrated in regions of hot and humid climate, with abundant vegetation and in altitudes lower than 1000 meters. According to [5], the life cycle of *D. hominis* has two well-defined stages. The nonparasitic stage corresponds to the soil pupation and adult flies in forest, and the parasitic stage corresponds to the entire development of larvae in the subcutaneous tissue of the host. The flies copulate in the first 24 hours after their emergence. Few hours after fertilization, the females begin to frequent the vicinity of cattle corrals, meeting several species of fly vectors. The deposition of their eggs is made during the flight in the lateral–ventral region of the vector after its capture and immobilization. The incubation period of eggs in the vector is of approximately eight days, and when this vector meets the host, the larvae break the eggs and penetrate through the hair follicles into the skin causing nodular myiasis. The larval period can go from 25 to 60 days. It is at night or early in the morning that mature larvae leave the host and go to the ground to pupate, avoiding the sun.

In Colombia, [6] observed higher prevalence of *D. hominis* in rainy season. [7] reported the occurrence of dermatobiosis throughout the year in Argentina, with infection peaks in rainy season, with warmer temperatures and higher humidity. [8] observed a higher incidence of infestation by larvae of *D. hominis* in the months of November and March, in São Paulo, Brazil, with decreased incidence until June. Larger infestations by warble were verified in March and April, in the state of Paraná (Brazil), with lower incidence in August and September, according to [9]. The authors linked the higher incidence of this parasitosis with rainy season. As [10] says, the highest prevalence of *D. hominis* during the rainy season is due to the better development conditions for the parasite, where a greater number of larvae can reach the pupal stage.

[11] described that the warble is distributed in approximately 20 states in Brazil, with higher abundance in Rio Grande do Sul, Santa Catarina, Paraná, Rio de Janeiro, Espirito Santo, Distrito Federal, and Goiás. The author mentions that the parasite does not occur in the states of Amapá, Rondônia, Ceara, Rio Grande do Norte, and Sergipe. According to the author, the soil conditions in these places do not offer conditions for the parasite to complete its life cycle. According to [4], *D. hominis* life cycle is complete in 80–150 days.

Observations related to seasonal variations in *D. hominis* in the city of Governador Valadares, Minas Gerais, made by [12] revealed that there is a positive correlation between parasitism by larvae *D. hominis*, relative air humidity and rainfall. However, no relationship was observed between ambient temperature and parasitism rates. Seasonality studies of the warble in cattle from the city of Guaíba, RS, mentioned by [13] have shown that in the warmer seasons of the year, that is, during the spring and summer, infestations happen with higher intensity. [14] in surveys conducted in Campo Grande – MS observed higher rates of warble infestation in periods of higher rainfall and higher relative humidity, with no positive correlation between ambience temperature and infestations in animals and also reported the presence of larvae throughout the study period with maximum amounts in March and May.

By studying the seasonal fluctuation of *D. hominis* in bovine skins coming from slaughter-houses, [15] observed that the highest percentages of infestation occurred when the months before had recorded increases in average temperature and rainfall. These factors may favor the penetration of larvae in the soil decreasing the pupation time of *D. homins* larvae. In addition, such climatic conditions also benefit its vectors' pupation.

In southeastern Brazil, the months of spring and summer, which correspond to the rainy season, are the most favorable period of year for the occurrence of dermatobiosis in cattle. Smaller infestations happen during the dry season in the months of autumn and winter according to [16] and [17].

According to [18], in Brazil, losses caused by of *D. hominis* larvae reach 250 million dollars per year.

2.2. Body distribution of *Dermatobia hominis* larvae

A study on variations related to infestations of cattle by *D. hominis* larvae was held in Viamão – RS by [19], when the author observed higher incidence of warble in the anterior left part of cattle. [13] observed that, in cattle, 73% of subcutaneous nodules caused by *D. hominis* larvae were distributed in the anterior parts. The most infected parts were the ribs (31.9% of the observed nodules), scapula (21.5% of the observed nodules), forelegs (17.8% of the observed nodules), and neck (8.8%).

[20] verified the parasite dynamics of warble, noting its incidence in relation to decubitus in cattle of the Canchin race, in São Carlos – SP. The incidence of parasitism was higher on the left side (14.2 nodules on average) compared to the right side (10.5 nodules). According to the author, the higher incidence of parasitism on the left side can be explained due to the fact that this region was more exposed to the vectors of *D. hominis'* eggs. In his observations it was possible to say that most of the animals during their rest leaned on their right side, that is, 2.360 animals observed, 1.183 had the habit of lying on their left side, while 1.447 were lying on their right side. In another study, [21] found that the regions of the forelimbs and the left blades were more parasitized. According to the author, low parasitism in posterior regions was due to the tail, which acts as a broom protecting such areas up to approximately the seventh rib. The data showed that, although protected posterior regions are equivalent to 41.06 % of the body surface of the animal, only 16.20 % were infested by warble. In another study by [22], in the city of Seropédica in the state of Rio de Janeiro, it was observed that the body region with the highest number of nodules was the blade, followed by the ribs and the forelimbs. It was also observed that, in cattle antimeres, the left side had 50.46 % of the nodules, and the right side 49.54 %. But this difference was not statistically significant. [23] conducted a study regarding the seasonal fluctuation of larvae *D. hominis* on cattle skins from slaughterhouses, observed a higher incidence of nodules caused by the larvae of *D. hominis* in the anterior region, with a 97.8 % rate.

[24] observed a significantly higher frequency of *D. hominis* nodules in females (16.7%) than in males (14.7 %). The presence of larvae in adult animals (15.4 %) is also more significant than in younger animals (12.1 %) and when it comes to the coat, the highest frequency of larvae was observed in the dark ones (black). Considering the body part, the one that was the most parasitized was the left anterior quadrant.

2.3. Organic dairy production system

In conventional livestock, the larva population on the cattle is controlled with the use of chemical larvicide; on the other hand, organic rural must meet the standards contained in the 60th Article of Normative Instruction No. 46, 2011, Ministry of Agriculture and Supply, which regulates organic production in Brazil, restricting the use of allopathic medicines [1]. The term "organic" refers to animal and vegetable food that are produced without the use of fertilizers;

pesticides; insecticides; antimicrobials; antiparasitic, transgenic, or any other drug that may contain harmful residues to human health, including agricultural products to conventional dairy farms [25].

Milk production in organic systems does not reach 0.1% of national production, which is about 25 million liters per year, due to several factors, such as: rural extension work enabling the process to small producers; the lack of scientific research adapting livestock production in organic system to the tropical reality; as well as food pasture fertilizers, racial patterns, and health care with the herd, such as endo- and ectoparasites control and mastitis [26].

3. Materials and methods

3.1. Location

The study was conducted from September 2009 to August 2010 in an area that belongs to the Sistema Integrado de Produção Agroecológica (Integrated Agroecological Production System) – SIPA (Fazendinha Agroecológica Km 47), technical cooperation project between Embrapa Agrobiologia, Empresa de Pesquisa Agropecuária do Estado do Rio de Janeiro (Agriculture Research Corporation of Rio de Janeiro State) (PESAGRO – Rio /Seropédica), and Universidade Federal Rural do Rio de Janeiro (Rural Federal University of Rio de Janeiro) [27]. SIPA is located in the city of Seropédica, metropolitan region of Rio de Janeiro state, currently occupying 70 hectares and incorporating, in addition to vegetable production area and fruits, a fragment of forest, a forest garden, and areas of agroforestry and ornamental species. Pastures subdivided into paddocks total 30 hectares.

3.2. Weather

The meteorological data used were temperature (T) of the air, relative humidity (RH), and precipitation (PP) obtained from the Agrometeorological station situated in SIPA's area.

The climate is hot and humid with little pronounced winter. The average temperature of the coldest month is higher than 20 ° C (68°F) and the maximum temperature in the summer can exceed 40 ° C (104° F). The rainfall is characterized by the existence of a rainy season in summer and dry in winter. The annual rainfall is around 1.300 mm, although it is mostly rainy in spring and summer, the occurrence of prolonged drought is common in the months of January and February [27].

3.3. Animals

The herd consisted of 40 crossbred dairy animals Zebu x European (Gir x Holstein), divided into lots of young and adult animals. The young ones were divided into two further lots: suckler calves (birth to 6 months) and weaned calves (from 7 months to 18 months or 330 kg), and a lot of adult animals consisting of dry cows, in lactation, and a bull. The determination of the coat of animals followed the Girolando characterization [28]. (Figures 1 A, B, C, and D).

Figure 1. Colors of the coat: **(A)** brown; **(B)** black on white; **(C)** red typical shades; **(D)** white on black **(E)** black.

3.4. Management of animals

The management system was semi-intensive: the animals remained in the corral during the day, where accumulation of manure could take place, and returned to the grass in the late afternoon. A physical model for organic milk production is implemented. Throughout the management, animal welfare, including avoidance of psychological stress in the herd, is prioritized. All the pickets have access to clean drinking fountains with good-quality water and shaded areas with afforestation. Containment fences are electrified and made with flat wire, in order not to represent a risk of injury to the animals.

The pastures are used in a rotation system. To supply the smaller forage production that happens in the dry period (period of lower growth of pastures), a cultivated area is managed to offer a forage supply in the trough. It is estimated that the period of lowest forage production in the region begins in mid-June and goes on until late October; that is, 135 days (or nine Fortnights) of drought and lower temperatures at night. A dairy Gir bull is used to ensure the reproduction of cows as well as the welfare of animals.

3.5. Health Management

The health management system established was developed for the SIPA project "Fazendinha Agroecológica Km 47." It is based on the folowing: animal welfare, strategic control of parasites, and homeopathic therapy, always stressing prevention as the most important aspect with regard to treatment. The specific objective was the reestablishment and maintenance of herd health in that organic system, and the general goal was to facilitate the structuring of an experimental organic dairy cattle system.

Homeopathic medicines have been prepared by the Pharmacy School from Instituto Hahne-manniano do Brasil. Drugs are in accordance with the rules of the Brazilian Pharmacy in the form of liquid presentation, and packaged in appropriate amber glass containers. The ways of administration are oral, nasal, or vaginal.

As already mentioned, throughout the management, the priority is the animal's welfare, including avoiding of psychological stress in the herd. "Good management practices in dairy cattle with emphasis on preventive health" established for this breeding system follow the definitions of the 60th Article of Normative Instruction No. 46, 2011, Ministry of Agriculture and Supply.

The basic requirements under Article 60 of MAPA IN No. 46 [1] are as follows: (1) follow the principles of animal welfare at all stages of the production process; (2) keep hygiene and health throughout the breeding process, consistent with current health legislation and the use of products that are authorized in organic production; (3) provide preventive health techniques; (4) offer nutritious healthy food, with quality and in correct amounts according to the nutritional requirements of each species; (5) offer good-quality water and in appropriated quantities, free of chemical and biological agents that may compromise their health and vigor, quality product and natural resources, according to the parameters specified by law; (6) the use sanitary facilities that are functional and comfortable; and (7) dispose in an environmentally appropriate way, the production wastes.

Vaccinations against FMD, brucellosis, clostridial diseases, salmonellosis, and rabies follow the current schedule in health-surveillance Ministry of Agriculture Livestock and Supply. Homeopathy is the adopted therapy for treatment and prevention of major diseases of dairy cattle, with a Homeopathic protocol developed for this creation system.

A supplement freely provided to the entire herd was formulated according to this system, composed of salt, sulfur (for animal feeding), and dicalcium phosphate.

3.6. Monitoring dermatobiosis (berne)

Inspection was performed biweekly (mapping the presence of larvae), totaling 915 inspections. The animals were inspected by anatomical demarcation, and their body divided into antimeres: anterior upper right (RADS), anterior lower right (RADI), posterior upper right region (RPDS), lower right posterior region (RPDI), anterior upper left (RAES), left anterior inferior (RAEI), posterior upper left region (RPES), and posterior lower left region (RPEI). The presence of the larvae (Figure 2) was observed in the different regions, and the data recorded in documents, according to the methodology of [29], with modifications (Figure 3).

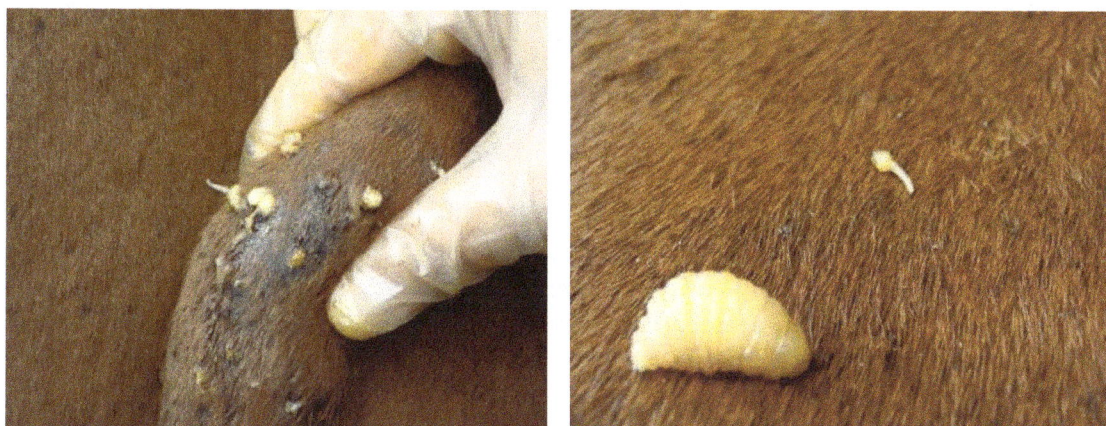

Figure 2. Presence of *D. hominis* larvae in subcutaneous tissue of cattle.

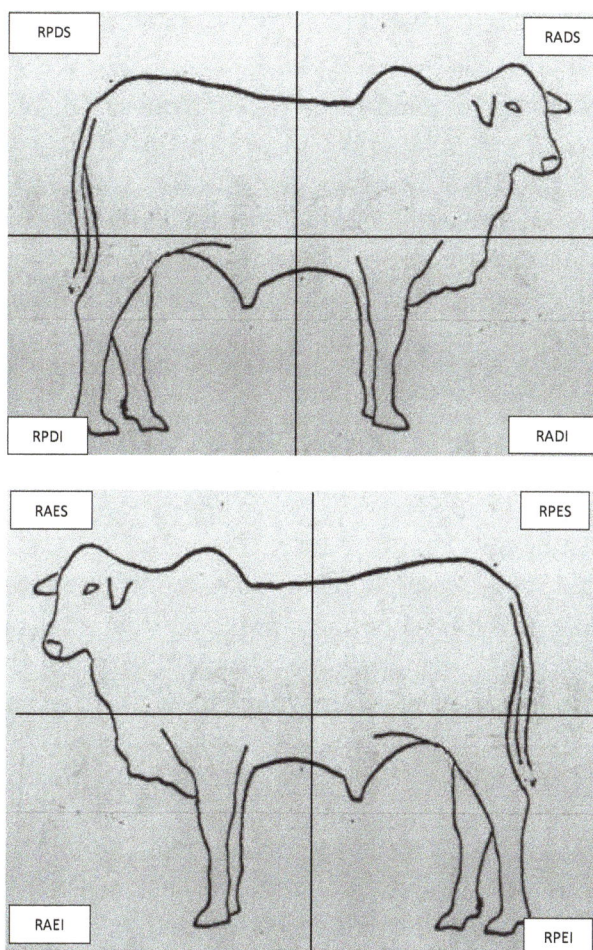

Figure 3. Field spreadsheet to map the dermatobiosis in cattle, according to the methodology of [29] with modifications.

3.7. Statistical Analysis

The berne description of the amounts into categories of each attribute were studied, and performed some exploratory data analysis through bar charts, box plots and calculating the average number of warble per studied animal. To compare the berne counts among the quadrants defined by anatomical demarcation, we used nonparametric Wilcoxon and Kruskal-Wallis test [30], due to the presence of nonnormal data [31]. To verify the association between the inherent variables to the animals and climate we used the generalized linear bivariate model of *Poisson* [32]. The dependent variable was the larva counted in each animal, the independent variables or explanatory variables were related to the animal profile (gender, age, and coat) and climatic factors (average temperature, rainfall, and relative humidity). As the dates of collection were different between adults and young animals (suckling calves and weaned calves), a stratified analysis was made taking into consideration the age of the animals involved in the study. The relative risk indicator is a measure of association, where two or more variables are correlated, being one of the ways used to the assessment in epidemiological statistics to answer the correlations between two outcome and exposure variables, where RR = 1 lack of association occurred; 0 <RR <1 protection factor, and RR> 1 risk factor.

In the period of study, 915 berne counts in cattle were made (inspections), in which 391 were in adult cattle, 356 in weaned calves, and 168 in suckling calves. Of the total, 784 females and 131 males were counted. Of the 915 counts, 354 were made in cattle coat with red color in typical shades, 180 in fur animals with white on black, 87 in cattle with black color coat, 198 counts in animal with light brown and dark coat, and 96 counts in animal with black on white coat.

To adjust the climate data to the study database, the average was calculated for each of them (Average temperature, rainfall, and relative humidity) taking into consideration a fifteen-day delay period preceding the collecting day.

All statistical analyses were performed using statistical package R [33].

4. Results and discussion

The results of monitoring of the herd dermatobiosis indicated that there was a significant predominance of parasitism in the females (total average 21.98 bernes per female against total average rating of 8.37 bernes per male), as shown in Figure 4 (A), where the average number of nodules per sex in each animal is observed. Also, greater variability in females than in males was observed, as shown in Figure 4 (B). Also in relation to gender, males showed a higher number of nodes on the right side (total of 4.46 against 3.90 on the left), where the RPDs (Posterior Right Upper Region) was the most infested (2.16). In females, the highest number of nodules were concentrated on the left side (total of 11.17 against 10.70 on the right) and RADS (Anterior Right Upper Region) was the most affected (6.98). Table 1 shows the average number of nodules per animal according to sex. It was found that there was significant difference (p-value <0.001) regarding the amount of bernes between males (8.37) and females (21.98).

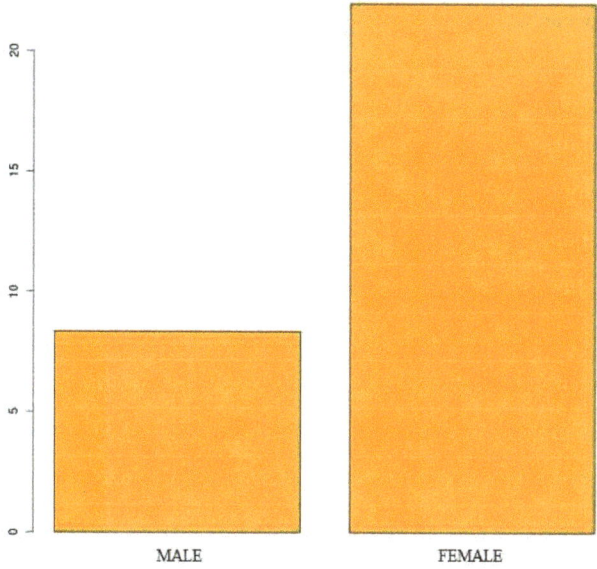

(a)

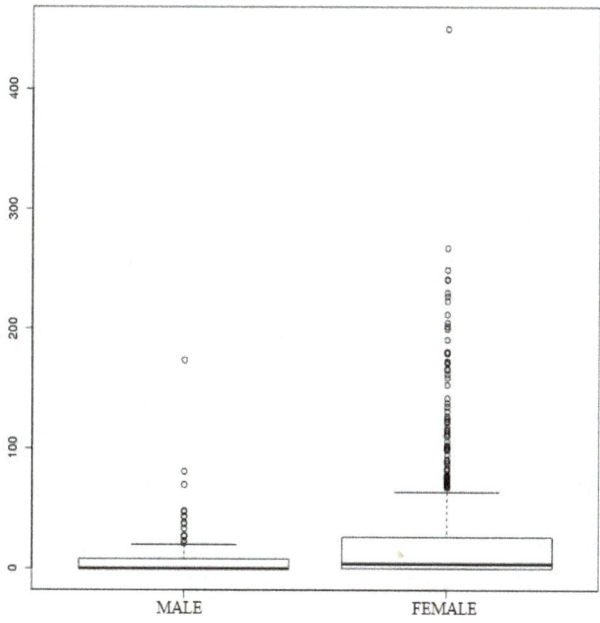

(b)

Note: Total of 915 counts of warble in cattle (inspections), with 784 females and 131 males in cattle. The average total number of bernes per female on the herd was 21.98 and the average total number of bernes per male was 8.37.

Figure 4. Distribution of parasitism in cattle according to gender **(A)**. Degree of infestation variability between sexes in the herd **(B)**.

Average number of warbles per animal		Gender		Wilcoxon P-value
		Male	Female	
	Animal Total	8.37	21.98	<0.001*
Left quadrant	Lower Anterior	0.59	1.55	0.004*
	Upper Posterior	1.44	2.58	0.017*
	Upper Anterior	1.75	6.46	<0.001*
	Lower Posterior	0.12	0.60	<0.001*
	Total	3.90	11.17	<0.001 *
Right quadrant	Lower Anterior	0.29	1.51	<0.001 *
	Upper Posterior	2.16	1.88	0.362
	Upper Anterior	1.89	6.98	<0.001 *
	Lower Posterior	0.12	0.55	0.0111 *
	Total	4,46	10,75	<0,001 *

* Significant values assuming a significance level of 5%

Table 1. Average number of bernes per animal according to sex and distribution in their respective quadrants

A significant prevalence of parasitism in females agrees with the results found by [24]. Regarding the most infected body region, there was divergence of results in other studies since all author studies cited here [19, 13, 21, 22, and 23] indicate a predominance of infestation in the anterior region, unlike the results found in males in this study, where the most affected body region was the posterior upper right region (RPDS) with an average of 2.16 bernes per animal. Also, in relation to the group of males in the herd, the prevalence of nodules on the right (total of 4.46 against 3.90 on the left) contradicts the results found by [19] in the study of Viamão – RS; as well as [20], who observed that the incidence on the left side is related to prevalence of the right lateral-sternal decubitus at rest time. [21] also found prevalence of parasitism on the left and [22] in a study conducted in Seropédica – RJ found no statistically significant difference between the number of nodules on the right and left sides of cattle.

Considering the age of the animals, it was found that the number of adult animals that were affected by berne (total of 31.55) was significantly higher (p <0,001) than younger animals of the herd (total of 8.0 in suckling calves and 12.21 in weaned calves); the variability in this group was also higher than the variability in the younger group. In the group of young animals, the most affected ones by the parasitosis were weaned calves and (total of 12.21 per animal), therefore, the group of suckler calves was the least infested by the larvae of *D. hominis* (total of 8.01 per animal) as shown in Figure 4 (A), where the average number of bernes per animal according to the age is observed. Figure 4 (B) shows the variability of the total number of bernes, considering the age of cattle. Based on age, in both groups of young animals, the most infested side was the left one (total of 4.04 and 6.47 per animal), and the most affected body part in the group of suckling calves was the RADS (Anterior Right Upper Region), averaging 1.69 berne per animal, while in the group of weaned calves it was the RAES (Left Anterior Upper Region), averaging 3.53 bernes per animal. In the group of adults, the more infested side was the left (total of 15.71 nodules per animal) and the most affected region was the RADS (Anterior Right Upper Region) (average of 10.68 bernes per animal), as shown in Table 2. It was found that there is significant difference (p <0.001) compared to the amount of bernes related to age.

Average number of grubs per animal		Age			Kruskal-Wallis p-value
		Suckling	Weaned	Adults	
	The animal Total	8.01	12.21	31.55	<0.001 *
Left quadrant	Lower Anterior	0.47	0.73	2.38	<0.001 *
	Upper Posterior	1.65	1.83	3.28	0.7614
	Upper Anterior	1.63	3.53	9.37	<0.001 *
	Lower Posterior	0.29	0.38	0.76	0.0033 *
	Total	4.04	6.47	15.71	<0.001 *
Right quadrant	Lower Anterior	0.40	0.78	2.19	<0.001 *
	Upper Posterior	1.69	1.38	2.48	0.6788
	Upper Anterior	1.69	3.23	10.68	<0.001 *
	Lower Posterior	12.26	12.43	0.62	0.1312
	Total	4.02	5.79	15.68	<0.001 *

* Significant values assuming a significance level of 5%

Table 2. Average number of berne per animal in the herd according to age and distribution in their respective quadrants

This study regarding the age of the animals, including the evaluation of results referring to the sides in which the highest level of infestations occurred, has shown in the youth group and adult group an agreement with results of previous researches. But, in the adults' group and in the suckling calves' group, although presenting a predominance of infestation on the left side (total of 15.71 and 4.02, respectively), it was observed that the most infested body region was the RADS (Anterior Superior Right Region), averaging 10.68 bernes per animal and RPDS (Posterior Superior Right Region), averaging 1.69 bernes per animal, respectively, different from that indicated in previous studies by [19] held in Viamão – RS, as well as [20], which linked the prevalence of parasitism on the left side to the right lateral-sternal decubitus at rest time. [21] also found prevalence of parasitism on the left and [22] in Seropédica – RJ did not find statistically significant difference between the number of nodules on the right and left sides of cattle.

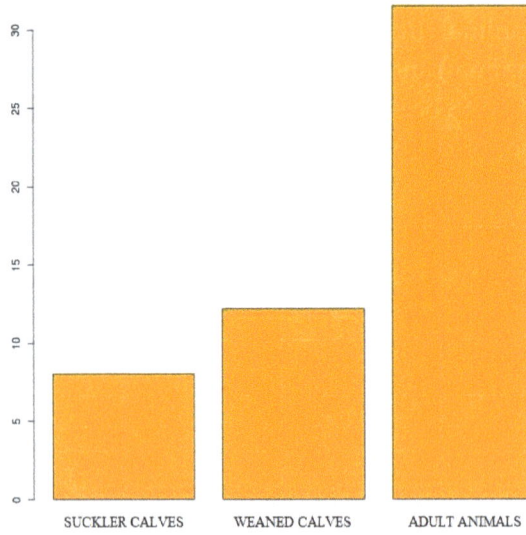

(a)

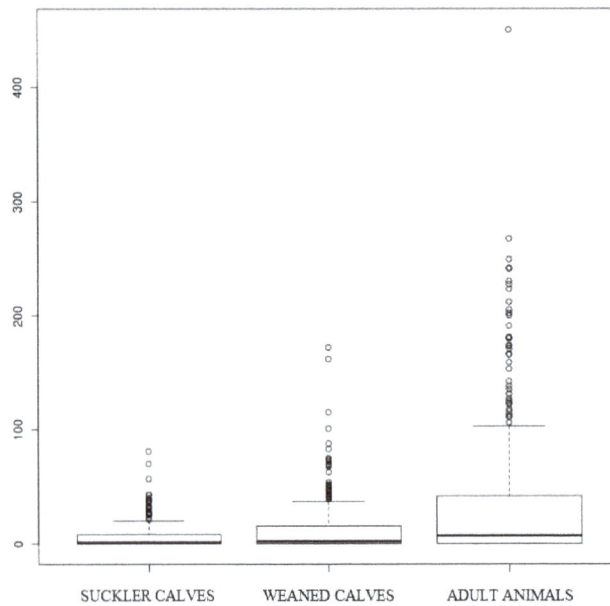

(b)

Note: Total of 915 bernes counts in cattle (inspections) in which 391 were in adult cattle, 356 in weaned calves, and 168 in suckling calves. The average total number of bernes per adult animal was 31.55, the average total number of bernes per weaned calf was 12.21, and the average total number of nodules per suckling calf was 8.01.

Figure 5. Average number of nodules per animal according to the age of the animal (A). Variability of the total number of bernes considering the age of animals **(B).**

Considering the presence of the larvae of *D. hominis* and the animal's coat, it is observed that the coat with a higher level of infestation is the black on white (total 36.69), while showing lesser infestation rates were typical red shades (14.13) and the light and dark brown (12.33).

In Table 3 we observe the average number of bernes per animal according to type of animal coat. There is a significant difference (p-value <0.001) comparing the amount of bernes between the coats. In Figure 5 (A) the average number of bernes per coat type in each animal is shown. Figure 5 (B) shows the variability of the total number of bernes in relation to the type of coat of the animal.

According to a study conducted by [24], the parasite frequency of occurrence was higher in dark-coat animals (black) unlike what was found in this study.

Average number of warbles per animal		Coat color					Kruskal-Wallis p-value
		White-Black	Red	Black-White	Black	Brown	
	Total per animal	17.08	14.13	36.69	29.82	12.33	<0.001 *
Left quadrant	Lower Anterior	1.80	0.86	3.37	1.59	0.31	<0.001 *
	Upper Posterior	1.05	1.70	5.04	2.33	1.98	<0.006 *
	Upper Anterior	4.30	4.72	10.69	8.74	2.57	<0.001 *
	Lower Posterior	0.44	0.38	0.91	0.80	0.40	0.0016 *
	Total	7.59	7.66	19.94	13.46	5.25	<0.001 *
Right quadrant	Lower Anterior	1.26	0.85	2.79	2.30	0.47	<0.001 *
	Upper Posterior	1.08	1.07	3.01	2.83	2.44	<0.007 *
	Upper Anterior	6.82	4.25	10.22	11.02	3.83	<0.001 *
	Lower Posterior	0.53	0.35	0.83	0.48	0.39	0.0892
	Total	9.59	6.52	16.52	16.08	7.07	<0.001 *

* Significant values assuming a significance level of 5%

Table 3. Average number of bernes per herd animal taking into consideration the coat type and distribution in their respective quadrants

As shown in Table 4, the months of highest occurrence of dermatobiosis were November and December, 2009, while the lowest levels of infestation by larvae of *D. hominis* were recorded in June and July, 2010. The period of highest infestation was the rainy season (spring and summer), and the record of the lower parasitism rates occurred during the dry season (fall and winter). The occurrence of parasites was observed throughout the study period. These findings coincide with observations of [6, 7, 10, 13, 16, and 17]. They also coincide with a study made by [14] on the observations of larvae presence throughout the study period, but differing in the months of maximum count. The results of this study also confirmed the observations of [8] in his study in the State of São Paulo, where he found higher occurrence of parasitosis in

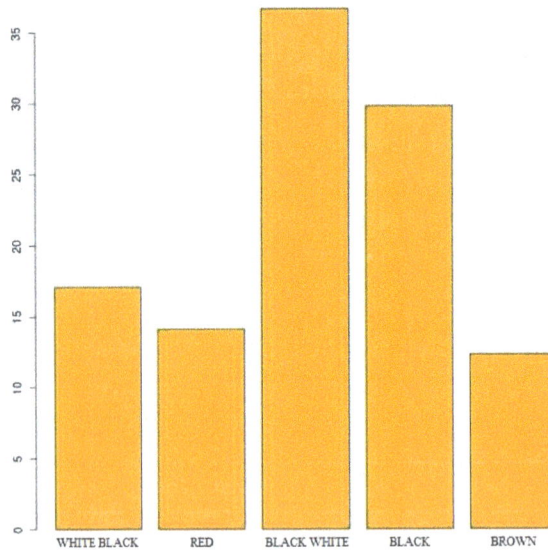

(a)

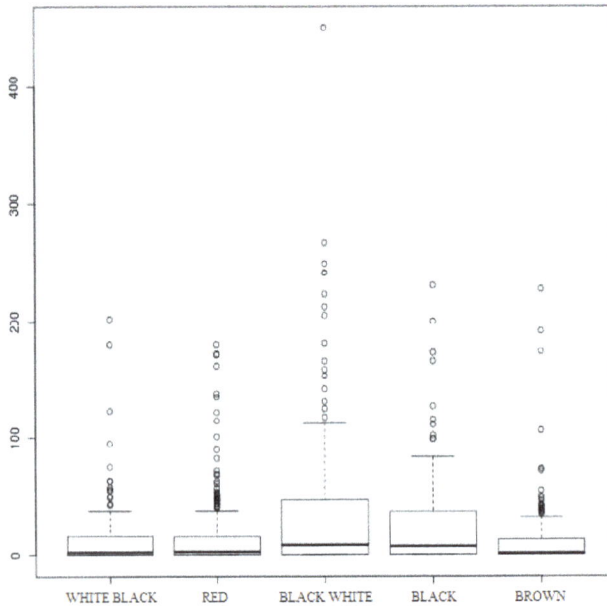

(b)

Note: Total 915 warble counts in cattle (inspections), of which 354 were in cattle with red coat color in typical shades, 180 in animals with white on black coat, 87 in black coat color cattle, 198 in animals with light brown and dark coat, and 96 in white on black coat color. The average total number of bernes per animal was 14.13 in cattle with red coat color in typical shades; 36.69 in animals with black on white coat; 29.82 in animals with a black coat color; 12.33 in cattle with light or dark brown coats; and 17.08 in animals with white on black coat.

Figure 6. Average number of bernes by coat type in each herd animal **(A)**. Variability of the total number of bernes in relation to the type of coat of animals **(B)**.

November, with a decrease until July, as well as the results found by [9] in his study in the state of Paraná, where major infestations in the rainy season period was also noted, differing only in the months of highest and lowest occurrence of dermatobiosis.

	Months/Years		Quantity of Bernes			
			Young animals		Adult animals	
			Average	Total	Average	Total
Rainy season	Spring	September/2009	26	1.381	63	2.263
		October/2009	21	821	52	1.886
		November/2009	35	1.422	99	3.546
	Summer	December/2009	33	1.393	104	3.748
		January/2010	5.3	223	20	709
		February/2010	3.1	125	11	409
Dry season	Autumn	March/2010	0.2	10	0.2	8
		April/2010	0.5	24	1,2	43
		May/2010	0.6	32	2.3	82
	Winter	June/2010	0	2	0.2	6
		July/2010	0.1	3	0.4	6
		August/2010	4	209	5.8	198

Table 4. Monthly averages of the average number of larvae *Dermatobia hominis* from September 2009 to August 2010

The fluctuation of the larvae of *D. hominis* during the studied period along with the climatic data are found in Figures 6 and 7, which show the highest levels of infestation occurring at the beginning of the study period as well as the highest rates of rainfalls, relative humidity air, and average temperature. The lowest averages in the occurrence of dermatobiosis occurred in the second half of the study period, coinciding with the lowest levels of rainfalls, relative humidity, and average temperature.

Risks relating to possible risk factors (intrinsic characteristics of the animal itself – gender, age coat, and climatic factors – rainfall, average temperature, and relative humidity) related to the occurrence of dermatobiosis are shown in Table 5. The results found, with reference to climatic variables, showed that with regard to rainfall and relative humidity, each increase of 1 mm^3 of water generates an average increase of 1.03 in the relative risk of occurrence of dermatobiosis in the herd, and each increase of 1°C in average temperature generates an average increase of 1.14 of relative risk to infestation by larvae of *D. hominis* in cattle. The results were significant for all studied weather variables. According to the study made by [15], increased percentages of infestation by larvae of *D. hominis* are related to the increase in average temperature and rainfall, which favors the penetration of larvae in the soil, reducing its time of pupation. Such

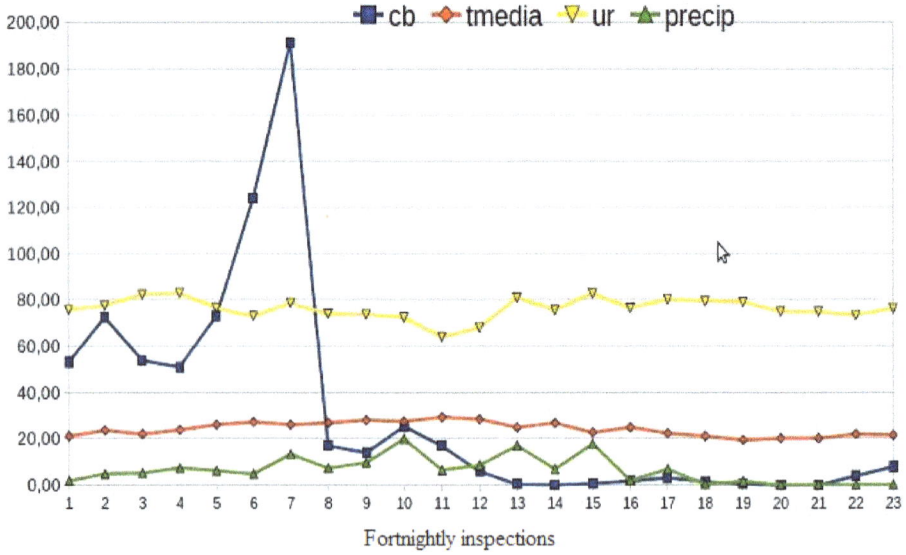

Note: The climatic variables are lagged by 15 days, so the average was calculated for each climate variable in the 15 days before each visit.

Figure 7. (cb) Average number of bernes in adult animals, (tmedia) average temperature, (ur) relative humidity, and (precip) rainfall during the experimental period.

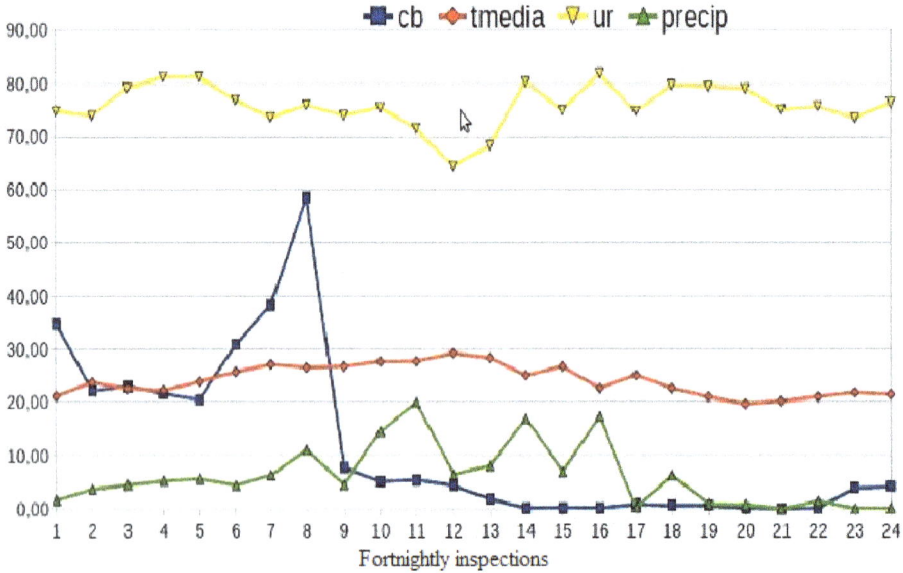

Note : The climatic variables are lagged by 15 days, so the average was calculated for each climate variable in the 15 days before each visit.

Figure 8. (cb) Average number of bernes in young animals, (tmedia) average temperature, (ur) relative humidity, and (precip) rainfall during the experimental period.

observations are confirmed by the results of this study, unlike the findings of [12, 5] in Governador Valadares, Minas, Gerais, and [14] in surveys conducted in Campo Grande – MS, which found no positive relationship between parasitism and average temperature.

Also, with regard to the results shown in Table 5, it was observed that adult females of the herd presented a relative risk 2.63 times higher than males to infestation by larvae of *D. hominis*.

Adult bovine animals had a relative risk for dermatobiosis 3.94 times higher than suckler calves, while weaned calves showed a relative risk of 1.52 times more than the suckling calves.

The black on white coats were the most susceptible to infestation by larvae of *D. hominis*.

The white on black coats showed a relative risk 2.98 times higher for developing dermatobiosis than light and dark-brown coated animals.

Variables	RR	IC 95%
Gender		
Male (ref.)	1.00	[1.00; 1.00]
Female	2.63*	[2.47; 2.79]
Age		
Suckler calves (ref.)	1.00	[1.00; 1.00]
Weaned calves	1.52*	[1.43; 1.62]
Adult animals	3.94	[3.72; 4.17]
Coat Color		
Brown (ref.)	1.00	[1.00; 1.00]
Black white	2.98*	[2.84; 3.12]
Black	2.42*	[2.29; 2.56]
White black	1.39*	[1.30; 1.48]
Red	1.15*	[1.09; 1.20]
Rainfall	1.03*	[1.02; 1.04]
Average temperature	1.14*	[1.13; 1.15]
Relative humidity	1.03*	[1.02; 1.04]

* Significant at 0.05 level

Table 5. Estimate of the relative risks (RR) and their respective confidence intervals of 95% (CI 95%) from the bivariate analysis of generalized linear models

The area occupied by Fazendinha Agroecológica Km47 incorporates a fragment of forest, a forest garden, and areas of agroforestry, and the climate is hot and humid with rainfall characterized by a rainy season in summer, and according to [4], the habitat of *Dermatobia hominis* is in hot and humid regions, with abundant vegetation and [5, 3, 2, and 14] state that there are plenty of those parasites on the margins of tropical forests and areas. It is noteworthy that the area or location of this study presents excellent conditions for the development of dermatobiosis, thus favoring the occurrence of high infestation levels as it was observed in the early months of the study. Also, because it is an organic system, the use of antiparasitic is strictly prohibited and contrary to the national law [1 and 25]; yet, this work observed improvement in the general appearance of the herd and improvement in conditions while handling of animals, as they have become extremely docile and receptive. Beyond these observations, the development of clinical diseases in cattle caused by parasitic load has not been registered.

5. Conclusions

- The ideal coats in this situation are the light and dark red and brown coated in typical shades;

- The degree of infestation was significantly higher in females than in males;

- It was not possible to say that there is influence of the right external–lateral decubitus in a parasitized body side;

- The temperature is the climatic factor that most influenced the parasitosis;

- The largest infestation rates occurred during the rainy season between spring and summer.

Author details

Mônica Mateus Florião[1*] and Wagner Tassinari[2*]

*Address all correspondence to: monicafloriao@hotmail.com

1 Rural Federal University of Rio de Janeiro - UFRRJ, Brazil

2 Department of Mathematics, Institute of Mathematical Sciences, UFRRJ, Rio de Janeiro, Brazil

References

[1] Brasil. Ministério da Agricultura, Pecuária e Abastecimento. Instrução Normativa n. 46 de 6 DE OUTUBRO DE 2011. Brasília, 2011.Disponível em: http://www.agricultura.gov.br/arq_editor/file/Desenvolvimento_Sustentavel/Organicos/Produtos%20Fitossanit%C3%A1rios/Home/IN_46_Prod_Animal_e_Vegetal_Organica-revoga_IN_64.pdf Acesso em 18 nov 1014.

[2] Neiva, A. Algumas informações sobre o berne. *Chac. Quint.*, v. 2, n. 1, p. 3-8, 1910.

[3] Andrade, E. N. Pesquisa sobre o berne, sua freqüência no homem, nos bovinos, suínos e eqüinos, e da aplicação de um novo método de provável eficiência para o seu combate. *Bolm. Biol. São Paulo..* v. 6, p.25 – 31, 1927.

[4] Moya Borja, G. E. Controle biológico do berne, *Dermatobia hominis* e de seus foréticos: crise e perspectiva. *Revista Brasileira de Parasitologia Veterinária*, v.13, supl. 1, p. 111-113, 2004

[5] Guimarães, J. H and Papavero, N. *Myiasis in Man and Animals in the Neotropical Region*. São Paulo: Editora Pleiade. 1999. 308 p.

[6] Mateus, G. *Dermatobia hominis* and its life cycle. *Inst. Colomb. Agropec.*, v. 2, p. 3-19, 1967.

[7] Lombardero, O. J. and Fontana, B. A. La ura (*Dermatobia hominis*) em la provincia de Formosa. *Gac. Vet. Buenos Aires*, v. 30, n. 215, p. 297-306, 1967.

[8] Lello, E.; Pinheiro, F. A. and Noce, O. F. Epidemiologia de miíases no Município de Botucatu (São Paulo). *Arq. Vet. U.F.M.G.*, v. 34, n. 1, p. 93-104, 1982.

[9] Magalhães, F. E. P. and Lesskiu, C. Efeito do controle do berne sobre o ganho de peso e qualidade dos couros em novilhos de corte. *Pesq. Agrop. Bras.*, v. 17, n. 2, p. 329-330, 1982.

[10] Moya Borja, G. E. O Berne: biologia, comportamento e controle. *Agroquímica*, v. 17, p. 19-26, 1982.

[11] Horn, S. C. O couro e seus problemas. *Bol. Def. San. Anim. Min. Agric.*, Brasil, p. 40, 1984.

[12] Maia, A. A. M. and Guimarães, M. P. Distribuição sazonal de larvas de *Dermatobia hominis* (Linnaeus Jr., 1781) (Diptera: Cuterebridae) em bovinos de corte da região de governador Valadares, Minas Gerais. *Arq. Bras. Méd. Vet. Zoot.*, v. 37, n. 5, p. 469-475, 1985.

[13] Bellato, V.; Paloshi, C. G.; Souza, A. P. De; Ramos, C. I. and Sartor, A. A. Variação sazonal das larvas da mosca do berne em bovinos no planalto catarinense. *Comunicado Técnico da EMPASC*, v. 101, p. 1 – 7, 1986.

[14] Gomes, A.; Koller, W. W. and Silva, R. L. *Haematobia irritans* (Díptera: Muscidae) as a carrier of *Dermatobia hominis* (Díptera: Cuterebridae) at Campo Grande, Mato Grosso do Sul State, Brazil. *Rev. Brás. Parasitologia. Vet.*, v. 7, n. 1, p. 69-70. 1996.

[15] Brito, L.G. and Moya Borja, G.E. Flutuação sazonal de *Dermatobia hominis* em peles bovinas oriundas de matadouro. *Pesquisa Veterinária Brasileira*, v. 20, n. 4, p. 151-154, 2000.

[16] Pinto, S. B. Valentim-Zabott, M.; Rochadelli, R.; Vendruscolo, E. C. G.; Fernandes, N. L.; Freitag, A. C.; Montanucci, C.; Lesskiu, P. E. and Spessatto, D. D. Eficácia de nú-

cleo homeopático na prevenção da infestação por *Dermatobia hominis* e *Haematobia irritans* em bovinos. *Arch. Vet. Sci.,* v. 10, n. 1, p. 57-62, 2005.

[17] Souza, F. S.; Fonseca, A. H.; Pereira, M. J. S.; Silva, J. X. and Goes, M. H. B. Geoprocessamento aplicado à observação da sazonalidade das larvas da mosca *Dermatobia hominis* no município de Seropédica – RJ. *Arquivo Brasileiro de Medicina Veterinária e Zootecnia,* v. 59, n. 4, p. 889-894, 2007.

[18] Grisi, L.; Massard, C. L.; Moya Borja, G. E. and Pereira, J. B. Impacto econômico das principais ectoparasitoses em bovinos no Brasil. *A Hora Veterinária,* v. 21, n. 125, p. 8-10, 2002.

[19] Oliveira, C. M. B. Variações mensais de infestações de bovinos por larvas de *Dermatobia hominis* (L. Jr.) em Viamão – RS. *Arq. Fac. Vet. U.F.R.G.S.,* v. 13, p. 64-64, 1985.

[20] Oliveira, G. P. Dinâmica parasitária de bernes em bovinos. Incidência em relação ao decúbito. *Pesq. Agrop. Bras.,* v. 26, n. 4, p. 467-471, 1991a.

[21] Oliveira, G. P. Parasites dynamics of *Dermatobia hominis* (L. Jr.), 1781) in cattle. II. Density, relationship between body regions and effects of the "brush." *Turrialba Publ.,* v. 41, n. 3, p. 359-366, 1991b.

[22] Maio, F. G.; Souza, W. M.; Grisi, L.; Sanavria, A. and Figueiredo, M. A. Distribuição sazonal das larvas de *Dermatobia hominis* (Linnaeus Jr., 1781) em bovinos leiteiros no município de Seropédica, Rio de Janeiro, Brasil. Revista Universidade Rural série Ciências da Vida, 21(½): 25-36, 1999.

[23] Brito, Luciana Gatto. Flutuação sazonal de *Dermatobia hominis* (Linnaeus Jr., 1781) (Díptera: Cuterebridae) através de peles de bovinos Recém Abatidos no matadouro do Município de Piraí – RJ e Infestação Artificial do Berne em Suínos e Eqüinos. 2000. 75 f. Tese (Doutorado em Parasitologia Veterinária) – Instituto de Biologia, Universidade Federal Rural do Rio de Janeiro, Seropédica.

[24] Sanavria, A. Cardoso, P. G. Morais, M. C. De and Barbosa, C. G. Distribuição e freqüência de larvas de *Dermatobia hominis* (Linnaeus Jr., 1781) (Diptera: Cuterebridae) em peles de bovinos. *Parasitologia Latinoamericana,* v. 57, n.1, p. 21-24, 2002.

[25] Instituto Biodinâmico. *Diretrizes.* 10ª ed. IBD, Botucatu, SP, 2000. 72 p.

[26] Alves, A.A. Panorama atual da produção orgânica de leite no Brasil. *Rev. Agroecol. Hoje.,* v. 29 p. 24-25, 2005.

[27] Almeida, D. L. De, Gerra, J. G. M. and Ribeiro, R. L. D. Sistema Integrado de Produção Agroecológica: uma experiência de pesquisa em agricultura orgânica. Seropédica, RJ, Documentos 169, Embrapa Agrobiologia, 37 p. 2003.

[28] Meldau D.G. Gado Girolando. Disponível em: <http://www.infoescola.com/pecuaria/gado-girolando/>. Acesso em: 10 Março, 2014.

[29] Costa H.N.A. and Freitas M.G. Efeito do composto fosforado DOWCO-14 (Narlene) administrado pela via oral, sobre a freqüência do berne em bovinos. *Arq. Fac. Vet. UFMG*, 13:145-156, 1961.

[30] Siegel, S. *Estatística Não-Paramétrica*. São Paulo: McGraw-Hill do Brasil, 1975. 352 p.

[31] Campos, H. *Estatística Experimental Não-Paramétrica*. 3ª ed. Piracicaba: FEALQ, 1979. 97 p.

[32] Nelder J.A. and Wedderburn R.W.M. Generalized Linear Models. *J.Royal Stat Soc.* A, v. 135, n. 3, p. 370-384, 1972.

[33] R Foundation for Statistical Computing, R a language and environment for statistical computing. R Development Core Team. Vienna, Austria, 2009. ISBN 3-900051-07-0, URL. Disponível em : http://www.R-project.org>. Acesso em: 20 Abril, 2014.

6

Organic Tuber Production is Promising — Implications of a Decade of Research in India

Suja Girija, Sreekumar Janardanan, Jyothi Alummoottil Narayanan and Santosh Mithra Velayudhan Santhakumari

Additional information is available at the end of the chapter

Abstract

Alternative soil management practices like organic farming assume significance in the context of climate change for safe food production. Yams (white yam, greater yam and lesser yam) and edible aroids (elephant foot yam (EFY), taro and tannia) are tuberous vegetables with good taste and nutritive value. Six field experiments were conducted at the ICAR-Central Tuber Crops Research Institute, Thiruvananthapuram, India, over a decade (2004–2015) to compare the varietal response, yield, quality and soil properties under organic vs conventional system and develop a learning system. The elite and local varieties of EFY and taro and the three species of yams, including trailing and dwarf genotypes, responded equally well to both the systems. Organic management enhanced the yield by 10–20% and the net profit by 20–40% over chemical farming. The tuber quality was improved with higher dry matter, starch, crude protein, K, Ca and Mg contents. The anti-nutritional factor in EFY, oxalate content, was lowered by 21%. Physico-chemical and biological properties of soil were favoured and the organic system scored a significantly higher soil quality index. The cost-effective technologies were field validated. A learning system developed using artificial neural networks predicted the performance of EFY organic production system.

Keywords: Eco-friendly farming, root crops, yield, quality, soil health, learning system

1. Introduction

Worldwide concerns regarding food safety, environmental degradation and threats to human health have aroused interest in alternative sustainable agricultural systems [1]. "Land degradation" is considered to be one of the world's greatest environmental challenges as per the UN millennium ecosystem assessment. Globally, 40% of the arable land is seriously degraded and

11% of this is situated in Asia [2, 3]. The land quality for food production ensures future peace. "Organic farming" is a viable option that enables sustainable production, maintenance of soil health, protection of human health and conservation of environment. It envisages non-use of synthetic chemicals, reduced use of purchased inputs and maximum use of on-farm-generated resources [3].

High input conventional agriculture that uses large quantities of chemical inputs and few C additions silently results in irrevocable ecological and environmental calamities [4, 5]. The necessity for environmental conservation along with the desire for safe foods has made organic farming one of the fastest growing agricultural enterprises [6]. It is well documented that there is a great demand for organic produce because of the belief that organic foods are more nutritious than conventionally grown ones [3, 7]. However, the nutritional or qualitative superiority of the organic food has yet to be proved conclusively. Reduced energy use and CO_2 emissions, employment generation, waste recycling and export promotion are the other merits of organic farming [3, 8, 9].

Tropical tuber crops constitute important staple or subsidiary food for about 500 million of the global population. Yams (*Dioscorea* spp.) and aroids are ethnic tuberous vegetables with good taste and medicinal values. They have high content of carbohydrate and are rich in energy. They also have higher protein content and better balance of amino acids than many other root and tuber crops. They are food security crops grown in tropical countries, mainly West Africa, the Caribbean, Pacific Islands and Southeast Asia. Tropical tuber crops in general and edible aroids like EFY, taro and tannia respond well to organic manures. Hence, there is great scope for organic production in these crops [3, 10–14]. There is a great demand for organically produced tuberous vegetables among affluent Asians and Africans living in Europe, USA and Middle East. Research and development on organic farming of tropical tuber crops is less focussed and documented. There is not much documented scientific evidence or information about the effects of organic management on yield, nutritional quality and soil health [3].

2. Why organic agriculture?

The major challenge faced by world agriculture is the production of food for a population of nine billion by 2050, with the anticipated climate change [15, 16]. There is an urgent call for transformations to increase the productive capacity and stability of smallholder agricultural production systems [15]. There is considerable discussion about the inadequacy of the present system of agricultural intensification and growth, which relies on increased use of capital inputs, such as fertilizers and pesticides [15, 17]. The generation of unacceptable levels of environmental damage and problems of economic feasibility are cited as key problems [17, 18]. Increasing concerns about the negative impacts of industrial agriculture have led to a serious debate over the feasibility of transition to alternative forms of agriculture, which are capable of providing a broad suite of ecosystem services while producing stable yields for human use [15}. Greater attention is thus being given to alternative models of intensification, and in particular, the potential of sustainable land management technologies. Such practices can provide private benefits for farmers, by improving soil fertility and structure, conserving soil and water, enhancing the activity and diversity of soil fauna, and strengthening the mechanisms of nutrient cycling [15]. These benefits can lead to increased productivity and stability of agricultural production systems [19–23] and offer a potentially important means of enhancing agricultural returns and food security as well as reducing the vulnerability of farming systems to climatic risk. Organic agriculture is one such promising alternative.

3. Organic farming feasible in selected areas and crops in India

In India, approximately 62% of cropped area is rain-fed, where there is little or no use of fertilizers and other agro-chemicals due to poor resources with smallholder farmers. Thus, promotion of organic farming in India is advocated initially in these rain-fed areas particularly in the hilly regions of northern and northeastern parts and dry land areas of the country. The Fertilizer Association of India has identified totally about 50 districts in the states of Orissa, Jharkhand, Uttranchal, Himachal Pradesh, Jammu and Kashmir, Rajasthan, Gujarat, Madhya Pradesh and Chhattisgarh as low-fertilizer-consuming districts with the consumption ranging from 1.79 kg ha^{-1} to 19.80 kg ha^{-1} as against the national average of 90.2 kg ha^{-1} [24, 25]. This means that there is immense scope for organic farming in these selected areas and for selected crops in India, like pulses, oilseeds, tuber crops, etc., for which conventionally little or no fertilizers and agro-chemicals are used. On the other hand, some areas growing tea, coffee, cashew, nuts and spices may be easily brought under organic farming with a thrust on export of organic produce. In other words, rather than promoting organic farming *en masse*, it would be appropriate to carefully delineate areas or crops, where fertilizer use is nil or nominal, or demarcate export-oriented crops that can give a reasonable yield of high-quality produce without using chemicals. It is noteworthy that tuber crops hold great promise in this regard [24].

4. Tuber crops: Underground crops with hidden treasures

Tropical tuber crops, including cassava, yams (greater yam, white yam and lesser yam), sweet potato and aroids (EFY, taro and tannia), form the most important staple or subsidiary food for about 500 million global population [24]. Tuber crops are the third most important food crops for humans after cereals and grain legumes. These crops possess high photosynthetic ability, have the capacity to yield under poor and marginal soil conditions and can tolerate adverse weather conditions. They are also recognized as the most efficient in converting solar energy, cassava producing 250×10^3 kcal ha^{-1} and sweet potato 240×10^3 kcal ha^{-1}, when compared with 176×10^3 kcal ha^{-1} for rice, 110×10^3 kcal ha^{-1} for wheat and 200×10^3 kcal ha^{-1} for maize; hence, the tropical root crops are known to be a cheap source of energy supply. They can serve as a substitute for cereals due to higher contents of carbohydrates and calories. The higher biological efficiency and the highest rate of dry matter production per unit area per unit time make tuber crops inevitable components of our food security systems. Besides, they have the potential to serve as sources of alcohol, starch, sago, liquid glucose, vitamin C and raw materials for many other industrial products and animal feed. At times of famine, tuber crops have come in handy to overcome catastrophes and provide relief from hunger [24].

Tuber crops are cultivated in India mainly as rain-fed crops in the southern, eastern and northeastern states. These crops are the source of livelihood to small and marginal farmers and tribal population in these areas. Cassava production is mainly reported in the states of Kerala, Tamil Nadu, Andhra Pradesh and NEH regions. Sweet potato is cultivated mainly in the states of Orissa, Bihar, Jharkhand, eastern Uttar Pradesh, West Bengal, Madhya Pradesh, Maharashtra and Karnataka. Other tuber crops like yams (greater yam, white yam and lesser yam) and aroids (EFY, taro and tannia), popular as vegetables, are not yet commercially cultivated, being confined only to the home gardens in almost all the states (except EFY, which is cultivated on a commercial scale in Andhra Pradesh) [24].

5. Prospects of organic farming in tropical tuber crops

Organic farming is a viable strategy targeting on sustainable production and soil, environmental and human health hand in hand. Conventional agriculture using chemical inputs results in higher yield, but it is ecologically unfriendly as it has negative impacts on food, soil, water and environmental quality. Indiscriminate use of chemical fertilizers for decades has lowered the organic carbon status of our soils to <1%. Moreover, pesticide residues cause concern over the safety of food. In traditional agriculture, though the use of chemicals (fertilizers and pesticides) is not in practice, adequate care is not often taken for the maintenance of soil health and fertility [24].

Most of the tuber crops are grown by small and marginal farmers in rain-fed areas and tribal pockets and hence the use of chemical fertilizers and insecticides is limited except in the case of cassava in the industrial production areas of Tamil Nadu (Salem, Dharmapuri, Namakkal, and South Arcot districts) and Andhra Pradesh (Rajahmundry district). Tuber crops in general and aroids in particular, like EFY, do respond well to organic manures and there is considerable scope for organic production in these crops. Further, the tropical tuber crops are well adapted to low-input agriculture. They are less prone to pest and disease infestations. Research work done in India and elsewhere had shown that the use of chemical fertilizers are beneficial in maximizing production of these groups of crops. A perusal of data in Table 1 indicates the organic production potential of tropical tubers and experimental evidences clearly indicate that productivity can be achieved satisfactorily even in the absence of chemical fertilizers through proper supplementation of nutrients using organic sources. Moreover, at present, there is a great demand for organically produced vegetables, particularly aroids and yams, among affluent Asians and Africans living in developed nations (Europe, USA and Middle East). The export of these tuberous vegetables will gain impetus through special government programmes like the Agri Export Zone (AEZ) Programme in Kerala [24].

Tuber crop	Tuber yield obtained due to application of organic manure (OM) alone		Tuber yield under OM + NPK			Reference
	OM used	Tuber yield (t ha⁻¹)	OM + NPK	Tuber yield (t ha⁻¹)	% increase or decrease over OM alone	
Cassava	FYM	10.45	FYM + NPK	28.17	+169.57	[26]
	Ash	12.25	FYM + NPK	28.17	+129.95	[26]
	Ash + FYM	13.29	FYM + NPK	28.17	+111.96	[26]
Sweet potato	FYM	15.57	FYM + NPK	18.88	+21.25	[27]
White yam (intercrop in coconut)	FYM	7.55	FYM + NPK	14.96	+98.15	[28]
	Coir pith compost	9.03	Coir pith compost + NPK	24.61	+172.53	[28]
	Green manuring with sunhemp	7.16	Green manure + NPK	16.06	+124.30	[28]

Source: Reference [29]

Table 1. Organic production potential of tropical tuber crops

6. Issues in organic tuber production

Practical applications and operational methodologies in organic farming, especially in tuber crops, are not available due to lack of comprehensive research in this field. Absence of package of practices recommendations for organic farming of tuber crops hinders the implementation and promotion of this sustainable alternative production system. Many methods and techniques of organic agriculture have originated from various traditional farming systems all over the world, where there is the non-use of chemical inputs. To the maximum extent possible, organic production systems rely on crop rotations, crop residues, animal manures, legumes, green manures, farm wastes, mineral-bearing rocks and aspects of biological pest control to maintain soil productivity, supply plant nutrients and control pests, diseases and weeds. Being highly responsive to organic manures and having fewer pests and disease problems when compared with cereals and vegetables, the main issue in organic production of tuber crops is the proper scientific use of a wide variety of cheaper and easily available organic sources of plant nutrients [24].

7. Strategies for organic tuber production

Building up of soil fertility of the land: Before the establishment of an organic management system, the fertility status of the land must be improved by growing green manure crops like cowpea twice or thrice in a year and incorporation of the green leaf matter at the appropriate pre-flowering stage. This will help re-establish the balance of the eco-system and offset the yield decline, if any, during the initial period of organic conversion, as tuber crops are highly nutrient-depleting crops. Virgin land or barren land, if available, will also be highly suitable for organic farming of tubers [24].

Use of planting materials produced by organic management: Varieties cultivated should be adapted to the soil and climatic conditions and as far as possible resistant to pests and diseases. Local market preference should also be taken into account. The planting materials should be produced by adopting organic management practices [24].

Meeting nutrient needs in organic tuber production: The potential organic sources of plant nutrients for tropical tuber crops are farmyard manure (FYM), poultry manure, composts like vermicompost, coir pith compost, mushroom spent compost, saw dust compost, press mud compost, green manures, crop residues, ash, oil cakes like neem cake, etc. Table 2 indicates the average nutrient contents in these organic sources [24].

Vermicompost, produced by chemical disintegration of organic matter by earthworms, is an ideal blend of plant nutrients with the worm enzyme and probiotics to boost the crop performance. It contains higher amount of nutrients, hormones and enzymes and has stimulatory effect on plant growth. If farmers can produce vermicompost utilizing on-farm wastes, organic farming of tuber crops becomes profitable [24].

Coir pith, an organic waste obtained as a by-product during the process of separation of fibre from coconut husk in the coir industry, is normally resistant to bio-degradation due to its high content of lignin, accumulating as an environmental pollutant. Extraction of 1 kg of coconut fibre generates 2 kg of coir pith, and in India, an estimated 5,00,000 MT of coir pith is produced per annum. The Coir Board in collaboration with TNAU has developed the technology for converting coir pith into organic manure using PITHPLUS, a spawn of edible mushroom, *Pleurotus sajor caju*. Coir pith compost developed from coir waste is a good form of organic manure and a soil conditioner and can be applied to tuber crops [24].

Organic manures	N (%)	P_2O_5 (%)	K_2O (%)
Farmyard manure	0.50	0.20	0.40
Poultry manure	1.20–1.50	1.40–1.80	0.80–0.90
Vermicompost	1.50	0.40	1.80
Coir pith compost	1.36	0.06	1.10
Press mud compost	1.30	2.20	0.50
Mushroom spent compost	1.84	0.69	1.19
Sawdust compost	1.00	0.50	0.50
Biogas slurry	1.41	0.92	0.84
Neem cake	5.00	1.00	1.50
Bone meal	3.50	21.00	–
Municipal compost	1.20	0.04	0.90

Source: Reference [24]

Table 2. Average nutrient contents of some organic manures

The practice of green manuring for improving soil fertility and supplying a part of N require-ment of crops is age old. Approximately 15–20 t ha^{-1} of green matter can be obtained from green manure crops like cowpea when grown in systems involving tuber crops. Nitrogen contribution by green manure crops varies from 60 to 280 kg ha^{-1} [24].

Biofertilizers offer a cheap and easily available source of nutrients, especially N and P, besides enhancing the efficiency of native and applied nutrients in the soil. The commonly used N biofertilizer for tuber crops is the N-fixing bacterium, *Azospirillum lipoferum*, which can partially meet the N demand of the crop. Powdered neem cakes also serve as an organic N source. These organic N supplements unlike the fertilizer N do not suffer much loss in the fields and enhances the N recovery. Phosphorus-solubilizing and phosphorus-mobilizing organisms such as phosphobacterium and mycorrhizae are helpful in augmenting P availa-bility of the soil [24].

Besides, natural reserves of rock phosphate are permitted for use as P fertilizer. Potassium for these crops can be supplied using K-rich organic amendments such as wood ash, rice straw and composted coir pith. K mobilizers can also be used for enhancing the K availability and meeting the K requirements. Harnessing the above-mentioned easily available organic sources of plant nutrients conjointly and judiciously to meet the nutrient needs of highly nutrient-exhausting crops like tropical tubers will definitely help maintain/promote productivity in organic farming in the absence of chemical inputs [24].

Pest, disease and weed management: When compared with cereals and vegetables, tuber crops have fewer pest and disease problems. Barring a few major ones, like cassava mosaic disease (CMD), cassava tuber rot, sweet potato weevil (SPW), *Phytophthora* leaf blight in taro, and collar rot in EFY, the others are of minor significance. In general, for the management of pests and diseases, non-chemical measures or preventive cultural techniques can be resorted to. This includes use of tolerant/resistant varieties, use of healthy and disease-free planting materials, strict field sanitation (against almost all), deep ploughing (e.g. tuber rot), roguing the field (e.g. CMD), use of pheromone traps (e.g. SPW), use of trap crops (e.g. SPW, root knot nematodes), adapted crop rotations, use of neem cake (collar rot, tuber rot), use of bio-control agents like *Trichoderma*, *Pseudomonas* (collar rot, leaf blight), etc. [24].

Normally, two hand weedings are advocated in tuber crops for efficient weed management. As most of the tuber crops (except sweet potato) take approximately 75–90 days for sufficient canopy coverage, raising a short-duration intercrop (like green manure/vegetable/grain cowpea, vegetables, groundnut, etc., in cassava, cowpea in yams and aroids) can also help to a great extent to reduce weed problem. Mulching the crop using any locally available plant materials (green leaves, dried leaves, etc.) immediately after planting (in yams and aroids) will help conserve moisture and regulate temperature, apart from weed control [24].

8. A decade of research on organic farming of tropical tuber crops

The following research programmes were taken up at ICAR-Central Tuber Crops Research Institute, Sreekariyam, Thiruvananthapuram, Kerala, India, during 2004–2015:

- Organic farming of EFY

- Varietal response to organic farming in EFY

- Validation and popularization of organic farming technology in EFY

- Organic farming of yams

- Organic farming of taro

- On-farm validation of organic farming of yams and taro

The major objectives were:

- To develop appropriate technologies for organic production of EFY, yams and taro, which would be safe and of good quality

- To assess the impact of organic farming in these crops on productivity, tuber quality, soil health and economics

8.1. Methodology

8.1.1. Study site, experimental design, treatments and test variety

Six separate field experiments were conducted at ICAR-Central Tuber Crops Research Institute, Thiruvananthapuram, India, over a decade (2004–2015) to compare organic management over conventional system in EFY, yams and taro in an acid Ultisol (pH: 4.3–5.0). The site experiences a typical humid tropical climate. The mean annual rainfall was 1,985 mm, maximum and minimum temperatures were 31.35°C and 24.50°C, respectively, and relative humidity was 76.65%. In general, for all the sites, prior to experimentation, the fertility status of the soil was found to be medium to high for organic C (0.75–1.03%), low for available N (159–255 kg ha⁻¹) and high for available P (142–217 kg ha⁻¹) and available K (337–528 kg ha⁻¹).

The impact of conventional, traditional, organic and biofertilizer production systems was evaluated in randomized block design (RBD) in EFY (var. Peerumade local) with five replications. Comparative response of five varieties of EFY (Gajendra, Sree Padma, Sree Athira and two locals) under organic and conventional farming was also evaluated in split plot design. The gross plot size was 4.5 m × 4.5 m (25 plants) accommodating nine net plants. All the three trailing genotypes of edible *Dioscorea* (white yam: *D. rotundata* (var. Sree Priya), greater yam: *D. alata* (var. Sree Keerthi) and lesser yam: *D. esculenta* (var. Sree Latha)) were evaluated under conventional, traditional and organic farming systems in split plot design. The gross plot size was 7.2 m × 3.6 m (32 plants of white yam and greater yam and 36 plants of lesser yam) accommodating 12 net plants of white yam and greater yam and 14 plants of lesser yam. The dwarf genotype of white yam (var. Sree Dhanya) was also evaluated under conventional, traditional, organic and integrated systems in RBD with five replications. Similarly, the response of three varieties of taro (Sree Kiran, Sree Rashmi and local) to conventional, traditional and organic farming systems was studied in split plot design. In split plot design, varieties/species were assigned to main plots and production systems to sub-plots and replicated thrice. Details of production systems are given in Table 3.

The on-station organic production technology developed for EFY was validated through on-farm trials (OFT) conducted at 10 sites covering 5 ha in Kerala under the project financed by the National Horticulture Mission. In yams and taro, the technologies were confirmed through OFT conducted at seven sites.

Chemical inputs were not used for a year prior to the start of the investigations. In "conventional plots", FYM + nitrogen, phosphorus, potassium (NPK) fertilizers were applied. Farmers' practice of using FYM and ash was followed in "traditional plots". In "organic farming plots", FYM, green manure, ash, neem cake and/or biofertilizers were applied to substitute chemical fertilizers. In "biofertilizer farming", FYM, mycorrhiza, *Azospirillum* and phosphobacterium were applied. In "integrated farming", FYM, chemical fertilizers and biofertilizers were used. Organically produced planting materials were used for the study.

Crop	Description of production systems			
	Conventional	**Traditional**	**Organic**	**Biofertilizers/integrated**
EFY	FYM @ 25 t ha^{-1} + NPK @ 100:50: 150 kg ha^{-1}	FYM @ 36 t ha^{-1} + ash @ 3 t ha^{-1}	Seed treatment in FYM + neem cake + *Trichoderma harzianum* slurry. Application of FYM @ 36 t ha^{-1} (FYM: neem cake mixture (10:1 ratio) incubated with *Trichoderma harzianum*) + *in situ* green manuring with cowpea (green matter @ 20–25 t ha^{-1}) + neem cake @ 1 t ha^{-1} + ash @ 3 t ha^{-1}	Biofertilizers FYM @ 25 t ha^{-1} mycorrhiza @ 5 kg ha^{-1} *Azospirillum* @ 3 kg ha^{-1} and phosphobacteria @ 2.5 kg ha^{-1}
Yams	FYM @ 10 t ha^{-1} + NPK @ 80:60:80 kg ha^{-1}	FYM @ 15 t ha^{-1} + ash @ 1.5 t ha^{-1}	FYM @ 15 t ha^{-1} + *in situ* green manuring with cowpea (green matter @ 15–20 t ha^{-1}) + neem cake @ 1 t ha^{-1} + ash @ 1.5 t ha^{-1} + biofertilizers (*Azospirillum* @ 3 kg ha^{-1} mycorrhiza @ 5 kg ha^{-1} and phosphobacteria @ 3 kg ha^{-1})	
Dwarf white yam	FYM @ 10 t ha^{-1} + NPK @ 80:60:80 kg ha^{-1}	FYM @ 15 t ha^{-1} + ash @ 1.5 t ha^{-1}	FYM @ 15 t ha^{-1} + *in situ* green manuring with cowpea (green matter @ 15–20 t ha^{-1}) + neem cake @ 1 t ha^{-1} + ash @ 1.5 t ha^{-1} + biofertilizers (*Azospirillum* @ 3 kg ha^{-1} and mycorrhiza @ 5 kg ha^{-1})	Integrated FYM @ 10 t ha^{-1} + NPK @ 40:30:80 kg ha^{-1} + biofertilizers (*Azospirillum* @ 3 kg ha^{-1} and mycorrhiza @ 5 kg ha^{-1})
Taro	FYM @ 12 t ha^{-1} + NPK @ 80:25:100 kg ha^{-1}	FYM @ 15 t ha^{-1} + ash @ 2.0 t ha^{-1}	FYM @ 15 t ha^{-1} + *in situ* green manuring with cowpea (green matter @ 15–20 t ha^{-1}) + neem cake @ 1 t ha^{-1} + ash @ 2.0 t ha^{-1} + biofertilizers (*Azospirillum* @ 3 kg ha^{-1}, mycorrhiza @ 5 kg ha^{-1} and phosphobacteria @ 3 kg ha^{-1})	

Table 3. Description of production systems in various organic farming experiments

8.1.2. Plant and soil measurements

Pooled analysis of yield data was performed. Yield stability index was calculated using the following formula: stability index = (Avg Y−SD)/Ymax, where Avg Y = average yield over five years, SD = standard deviation, Ymax = maximum yield over the five years. A stability index value towards unity indicates greater stability. Proximate analyses of tubers for dry matter, starch, total sugars, reducing sugars, crude protein, oxalates and total phenols [30–33], mineral composition of corms, namely P, K, Ca, Mg, Cu, Zn, Mn and Fe contents [34], chemical parameters of soil, namely organic C (soil organic matter (SOM)), pH, available N, P, K, Ca, Mg, Cu, Zn, Mn and Fe status [35], physical characters of the soil such as bulk density, particle density, water-holding capacity (WHC) and porosity [36], plate count of soil microbes, namely bacteria, fungi, actinomycetes, N fixers and P solubilizers [37] and the activity of dehydrogenase enzyme [38], were determined by standard procedures. Economic analysis was performed; net income and benefit:cost ratio were computed. The soil quality index (SQI) was computed in EFY based on the method developed by Karlen and Stott [39]. The analysis of variance of data was performed using reference [40] by applying analysis of variance technique (ANOVA) for RBD and split plot design.

8.1.3. Development of a learning system

A learning system was developed using artificial neural networks (ANN) to predict the performance of EFY production system [41, 42]. A three-layered system with one input layer, one output layer and one hidden layer was developed. The input layer neurons included temperature, rainfall, planting material, FYM, potassium, phosphorus, ash, neem cake, *Azospirillum*, phosphobacteria, mycorrhiza and green manure. The output layer neurons were total biomass, corm yield, canopy spread and plant height.

8.2. Implications

8.2.1. Varietal response to organic management

Pooled analysis indicated that the elite and local varieties of EFY and taro and all the three species of *Dioscorea* were on a par under both the systems (Figure 1). However, the Gajendra variety of EFY and all the species of *Dioscorea* yielded more under organic farming than conventional practice (Figure 1). In taro, all the varieties produced slightly higher yield under chemical farming.

8.2.2. Yield and economics

Organic farming resulted in 10–20% higher yield in EFY, white yam, greater yam, lesser yam and dwarf white yam, i.e., 20, 9, 11, 7 and 9%, respectively (Table 4). This is contrary to some of the reports that crop yields under organic management are 20–40% lower than those under comparable conventional systems [43, 44]. Taro preferred chemical-based farming as a slight reduction in the crop yield was noticed under organic farming (5%).

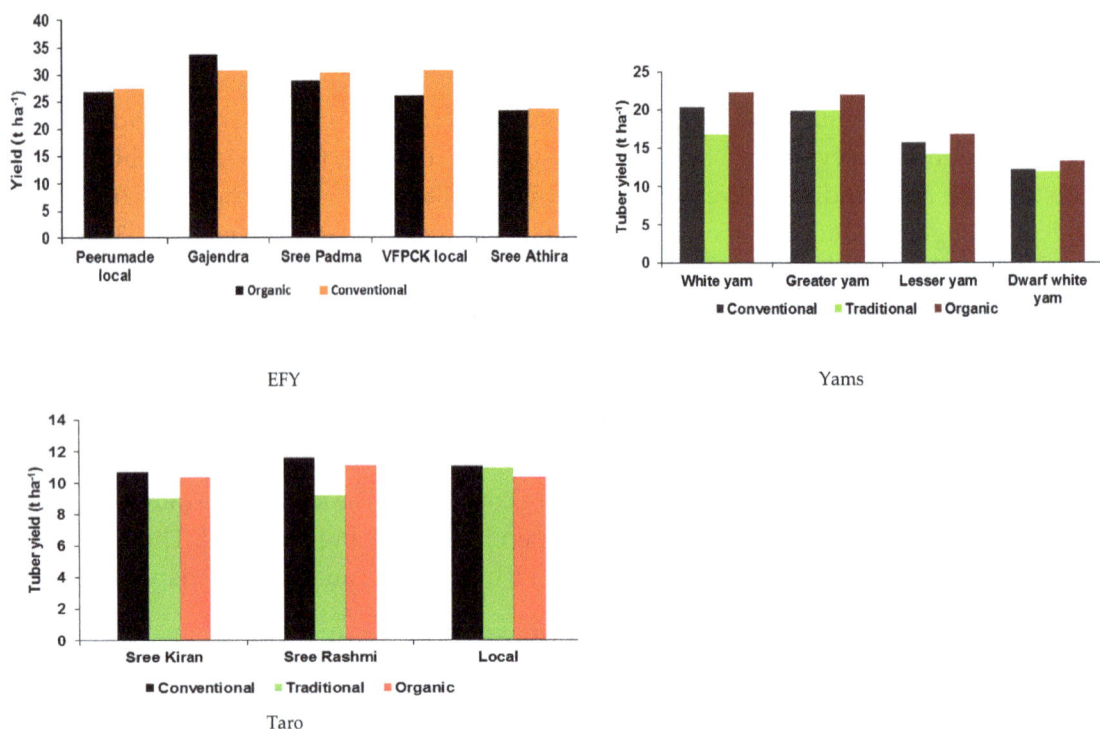

Figure 1. Varietal response to organic farming in tuber crops

It has been reported that yields were directly related to the intensity of farming in the prevailing conventional system [45, 46]. This means that in areas of intensive farming system, shifting to organic agriculture decreases the yield depending on the intensity of external input use before conversion [48, 49]. As EFY and yams are traditionally grown with low external inputs using organic wastes and manures available in the homesteads, organic management in the present study has shown a potential to increase yields over conventional practice. The higher yield may be due to the overall improvement in the physico-chemical and biological properties of soil under the influence of organic manures [9, 50, 51].

Tuber crop	Conventional	Organic	% increase/decrease
EFY	47.61	57.10	19.93
White yam	20.31	22.21	9.35
Greater yam	19.87	21.96	10.51
Lesser yam	15.75	16.83	6.85
Dwarf white yam	13.23	12.18	8.62
Taro	11.12	10.61	−4.58

Source: Reference [47]

Table 4. Yield (t ha^{-1}) under organic vs conventional management in tuber crops (pooled mean)

Tropical tuber crops, like EFY and yams, are nutrient-exhausting crops. In general, the nutrient removal by these crops yielding 17–33 tonnes of tuber was 112–180 kg N, 15–24 kg P and 93–239 kg K per ha [52]. The potential yield of these crops can be obtained by proper renewal of soil with adequate amounts of nutrients. These results highlight that in the absence of chemical fertilizers, in organic agriculture, a higher yield can be obtained through proper addition of nutrients based on soil testing by way of cheaper and easily available, on-farm-generated organic sources [3].

The long-term performance of organic vs conventional management in aroids and yams was analysed through the stability index calculated over a five-year period, and it was found that organic farming was equally stable as that of conventional practice (Figure 2).

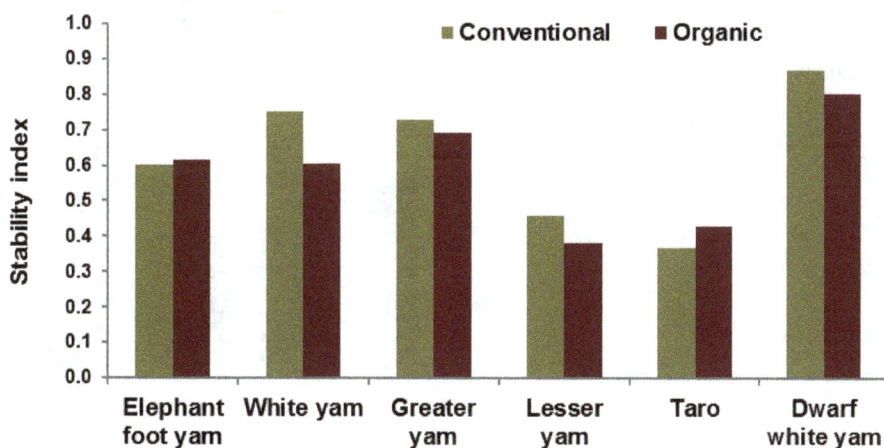

Figure 2. Yield stability index in organic vs conventional management in aroids and yams

The view of field experimentation in EFY is given in Figure 3. Yield trend over five years and pooled analysis indicated the significantly superior performance of organic farming in EFY (Figure 4; Table 5). Cost–benefit analysis in EFY indicated that the net profit was 28% higher and an additional income of Rs. 47,716 ha^{-1} was obtained due to organic farming, which was obviously due to 20% higher yield [12] (Table 5).

In yams, up to third year, organic farming proved to be superior; thereafter, it was on a par and slightly lower than conventional practice. Pooled analysis in yams indicated that organic farming was significantly superior to conventional practice and produced 9.12% higher yield (Figure 5; Table 6). Species × production systems interaction was absent. However, in all the species, organic farming produced slightly higher yield than conventional practice. Dwarf white yam also responded similarly to both the systems with slightly higher yield under organic practice (Figures 6 and 7).

In taro, yield trend over five years (except during the first year, when organic farming was superior to conventional practice) and pooled mean indicated that organic farming was on a par with conventional practice, but chemical farming produced a slightly higher yield (Table 4; Figures 8 and 9). This was because taro leaf blight could not be controlled by organic measures.

Field view of organic farming of EFY

Organic corms

Conventional corms

Bumper yield of organic corms

Figure 3. View of field experimentation in EFY

Production systems	Mean corm weight (kg plant⁻¹)	Corm yield (t ha⁻¹)	Gross income (Rs. ha⁻¹)	Gross costs (Rs. ha⁻¹)	Net income (Rs. ha⁻¹)	B:C ratio
	(Pooled mean of 5 years)					
Conventional	3.91	47.61	3,80,872	2,12,812	1,68,060	1.79
Traditional	3.69	44.96	3,59,680	2,18,800	1,40,880	1.64
Organic	4.69	57.10	4,56,776	2,41,000	2,15,776	1.90
Biofertilizers	3.45	42.07	3,36,528	2,16,240	1,20,288	1.56
CD (0.05)	0.292	3.550				

Source: Reference [12]

Table 5. Yield and economic advantage of organic farming over other production systems in EFY

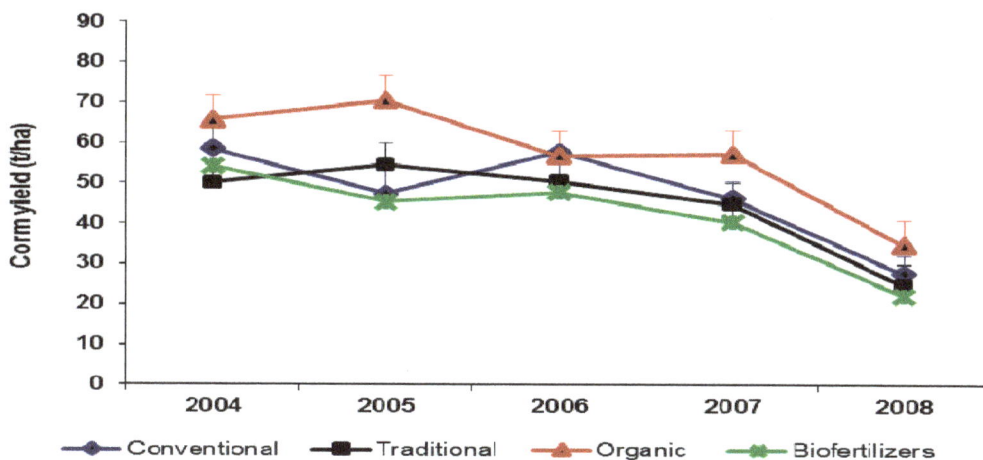

Figure 4. Yield trend over years as influenced by production systems in EFY

Field view of organic farming of yams: Green manuring, cost effective component

Organic white yam tubers   Organic greater yam tubers   Organic lesser yam tubers

Figure 5. Field experimentation on organic farming of trailing genotypes of yams

Species/production systems	Conventional system	Traditional system	Organic system	Mean of *Dioscorea* species
Dioscorea rotundata	20.31	16.76	22.21	19.76
Dioscorea alata	19.87	19.97	21.96	20.61
Dioscorea esculenta	15.75	14.18	16.83	15.58
Mean of production systems	18.64	16.97	20.34	
CD (0.05)	*Dioscorea* species: 1.23; production systems: 1.23; species × systems: NS			
Source: Reference [3]				

Table 6. Yield response of *Dioscorea* species to production systems (t ha^{-1}) (pooled mean)

Field view of organic farming of dwarf white yam: Green manuring, cost effective component

Figure 6. Field experimentation on organic dwarf white yam production

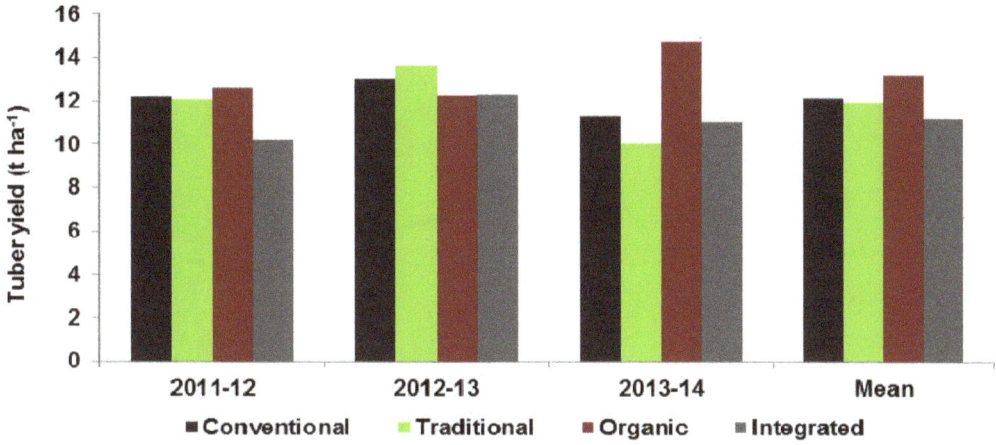

Figure 7. Yield trend over years as affected by production systems in dwarf white yam

Figure 8. Field view of organic taro production with green manuring as the component

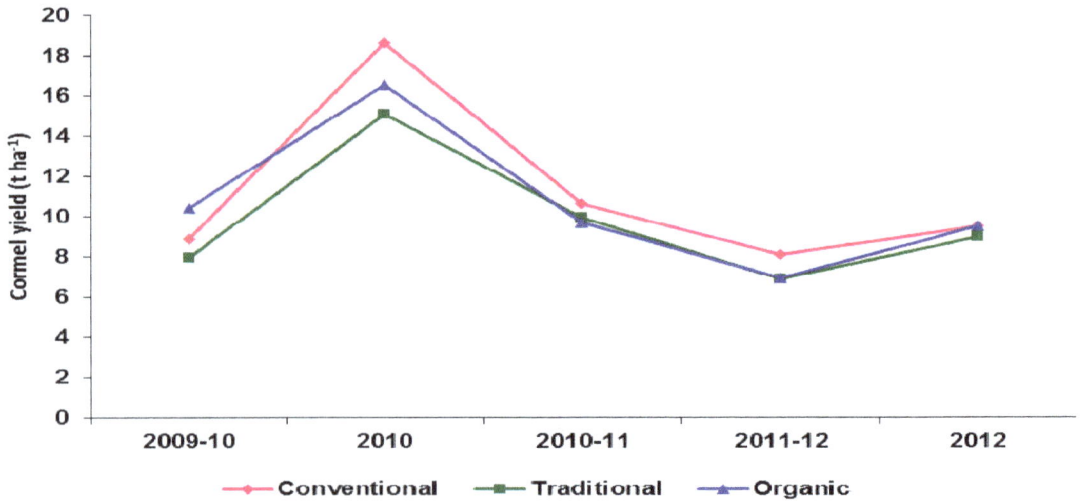

Figure 9. Yield trend as affected by production systems in taro

8.2.3. Nutritional quality of tubers

It is well known that plants absorb nutrients in the form of inorganic ions irrespective of whether the nutrient source is organic or inorganic. The absorbed nutrients are re-synthesized into compounds that determine the quality of the produce, which is largely decided by the genetic make-up of the plants [5, 12]. However, in the present research, dry matter and starch contents of organically produced EFY corms were significantly higher (by 7 and 13%), and crude protein (by 12%), K, Ca and Mg (by 3–7%) were slightly higher than those of conventional corms (Tables 7 and 8 and Figures 10 and 11). The anti-nutritional factor, oxalate, content in EFY was significantly lower (by 21%) due to organic management. Total sugar and total phenol contents of conventional corms were significantly higher. In yams, the tuber quality was improved with significantly higher Ca, slightly higher dry matter, crude protein (by 6–7%), K and Mg contents. Synthetic fertilizers enhanced the total sugars, reducing sugars and total phenol contents slightly. The cooking quality of organically produced tubers did not differ from that of conventional tubers (Tables 7 and 8 and Figures 10 and 11).

Earlier reports indicate that organic crops contain more dry matter, minerals, especially Fe, Mg and P, by 21, 29 and 14% over conventionally produced ones [7]. As stated in references [3, 53], higher levels of K were found in organic tomatoes. There is a higher population of micro-organisms in organically managed soil. These micro-organisms produce many compounds that combine with soil minerals and make them more available to plant roots [54], which might have ultimately enhanced the mineral content of tubers.

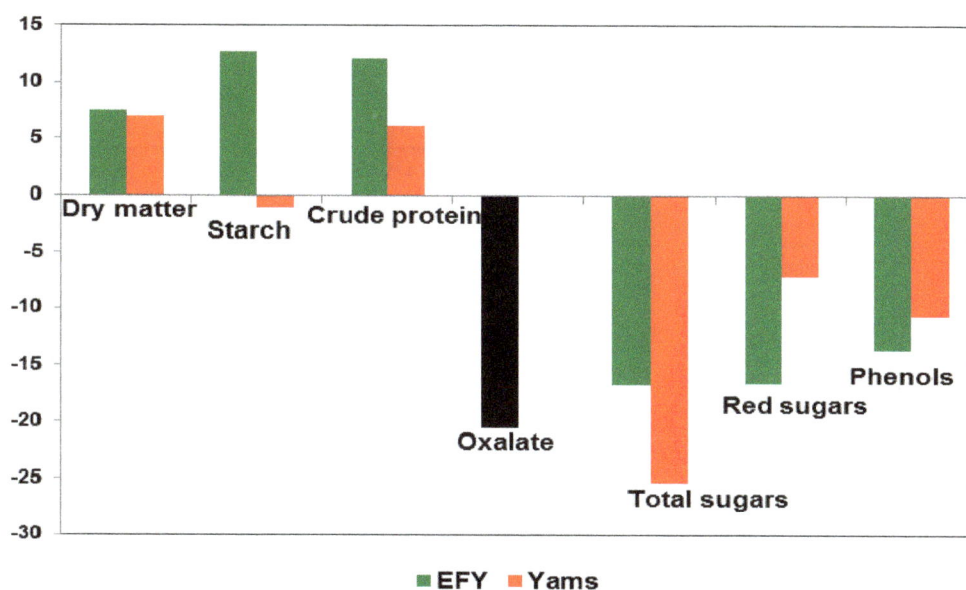

Figure 10. Per cent increase/decrease in biochemical parameters of organic tubers

Biochemical parameters	EFY			Yams		
	Organic	Conventional	CD (0.05)	Organic	Conventional	CD (0.05)
Dry matter (%)	21.41	19.93	1.061	33.56	31.36	NS
Starch (% FW basis)	16.54	14.68	0.937	26.40	26.70	NS
Crude protein (% FW basis)	2.04	1.82	NS	2.04	1.92	NS
Oxalate (% DW basis)	0.186	0.234	0.0259			
Total sugars (% FW basis)	1.98	2.38	0.257	1.88	2.52	NS
Reducing sugar (% FW basis)	0.65	0.78	NS	0.12	0.13	NS
Total phenols (mg 100 g^{-1})	69.70	80.80	8.28	37.20	61.60	NS

Source: Reference [14]

Table 7. Comparison of biochemical constituents of organic vs conventional tubers

Mineral content	EFY			Yams		
(mg 100 g⁻¹) (DW basis)	Organic	Conventional	CD (0.05)	Organic	Conventional	CD (0.05)
P	427.50	455.20	NS	411.80	472.40	39.79
K	1813.00	1714,00	207.40	1,051.30	1,026.70	NS
Ca	152.20	142.00	17.58	72.70	57.70	11.35
Mg	276.50	268.10	NS	180.60	161.70	NS
Cu	1.04	1.08	NS	0.24	0.29	NS
Zn	11.02	11.62	NS	4.49	4.45	NS
Mn	2.32	3.21	0.419	0.35	0.32	NS
Fe	71.90	86.60	NS	5.03	5.13	NS

Table 8. Comparison of mineral content of organic vs conventional tubers

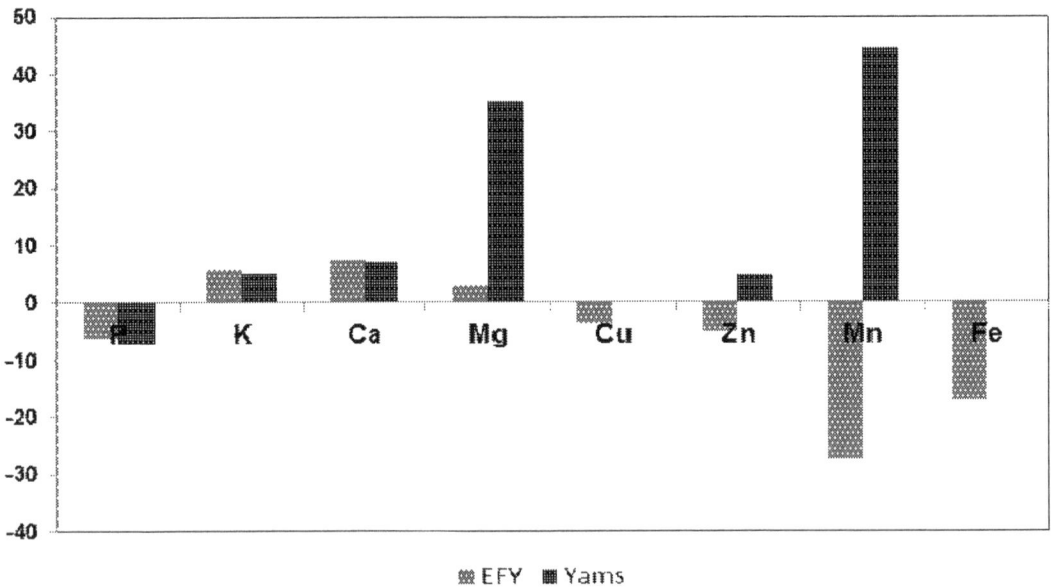

Figure 11. Per cent increase/decrease in mineral composition of organically produced tubers

Biochemical parameters of tubers were not significantly affected in taro and dwarf white yam. However, in taro, organic cormels had higher dry matter, starch and total sugars; conventional cormels had higher phenol, fibre and ash contents. Mineral content of cormels of taro also remained unaffected due to the production systems, though there was a slight increase in P, K, Ca and Mg contents in organic cormels (Figure 12).

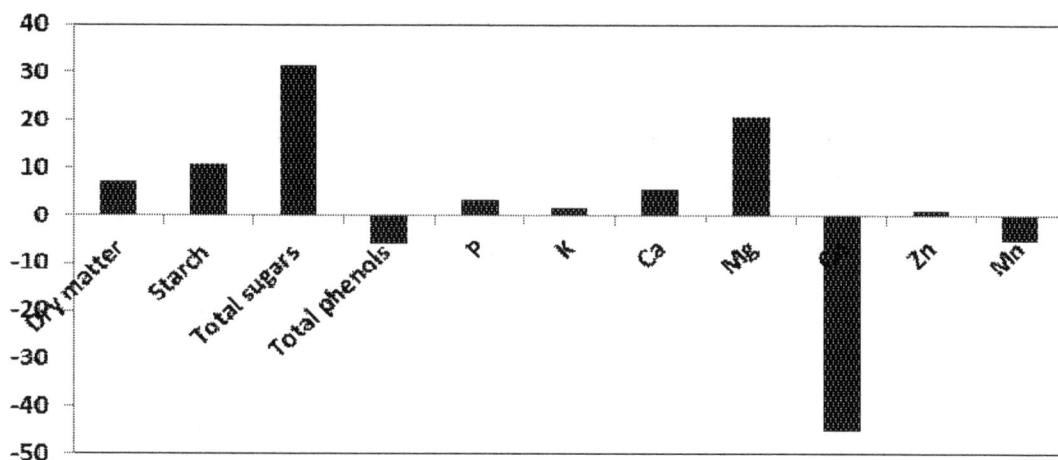

Figure 12. Per cent increase/decrease in biochemical and mineral composition in organic cormels of taro

8.2.4. Soil quality

8.2.4.1. Physico-chemical–biological indicators

The water-holding capacity was significantly higher under organic management (14 g cm^{-3}) in EFY and yams over conventional practice (11–12 g cm^{-3}). It was 28, 15 and 19% higher than that of conventional practice in EFY, yams and taro, respectively (Tables 9 and 10). Increased aeration, porosity and water-holding capacity of soils have been observed under organic management [51, 55, 56]. Moreover, changes in organic matter contribute to changes in soil biological and physical properties [9]. The higher organic C and organic matter contents under organic management in these crops might have resulted in the formation of stable soil aggregates leading to a slight decrease in bulk density and increase in water-holding capacity [3].

There was significant improvement in pH in organic farming (0.77, 0.46, 1.11 and 1.20 unit increase over conventional system) in EFY, trailing yams, dwarf white yam and taro (Tables 11 and 12). Several earlier workers have reported that significant improvement in pH under organic management may be due to elimination of NH_4 fertilizers, addition of cations especially via green manure applications, decrease in the activity of exchangeable Al^{3+} ions in soil solution due to chelation by organic molecules and self-liming effect of the Ca content in FYM (0.14%) and ash (20–40%) [3, 57–59].

The organic C content increased by 14–40% in organic plots over conventional plots in these crops (Tables 11 and 12). Higher organic C status of organic plots might be attributed to considerable addition of organic manures particularly green manure cowpea. In EFY, exchangeable Mg, available Cu, Mn and Fe contents were significantly higher in organic plots (Figure 13). Organic plots showed significantly higher available K (by 34%) in yams and available P in taro (Tables 11 and 12). Higher available P in organic plots may be due to solubilization of native P by organic acids during decomposition of organic manures and increased mineralization of P from the added organic manures [3, 12]. The higher content of available K in organic plots may be due to the higher content of K in the organic manures, especially green manure and ash (Table 2), greater mining of K from the sub-surface layers by the extensive root system of green manure crop of cowpea, and dissolution of K from the inaccessible K minerals in the soil by organic acids during green manure decomposition [3, 12].

The soil pH is the most important determinant of soil nutrient availability. As reported in reference [59], the rise in soil pH to neutral range under organic management in these crops might have enhanced the availability of major, secondary and micro-nutrients to some extent. Moreover, organic manures used in the study, FYM, green manure cowpea and neem cake that contain major, secondary and micro-nutrients might also have contributed to this [3, 12].

hysical parameters	EFY				Yams			
	Organic	Conventional	CD (0.05)	% increase or decrease	Organic	Conventional	CD (0.05)	% increase or decrease
Bulk density (g cm^{-3})	1.54	1.58	NS	−2.29	1.61	1.63	NS	−1.23
Particle density (g cm^{-3})	2.29	2.30	NS	−0.61	2.27	2.40	NS	−5.42
Water-holding capacity (%)	14.11	10.99	2.442	+28.38	14.21	12.38	1.604	+14.78
Porosity (%)	36.51	31.35	NS	+16.45	31.30	32.07	NS	−2.40

Source: Reference [14]

Table 9. Comparison of physical parameters of soil under organic vs conventional management in EFY and yams

At present, deficiency of secondary and micro-nutrients (Zn, S, B, Mo, Fe, Mn and Cu) is a rampant soil problem affecting crop productivity and profitability of farming in India [5, 12]. This is mainly due to the continuous use of high analysis fertilizers, which do not provide secondary and micro-nutrients. Based on research conducted for a decade in these crops, it has been proved beyond doubt that organic farming helps to reinstate soil productivity. Organic agriculture that envisages elimination of synthetic chemical fertilizers through strict use of organic manures helps to refurbish the soil health, by improving organic matter, neutralizing soil acidity, supplying almost all essential nutrients in the available form and ultimately conserving soil fertility [3, 5, 12].

Physical parameters	Taro			
	Organic	Conventional	CD (0.05)	% increase or decrease
Bulk density (g cm^{-3})	1.72	1.74	NS	−1.38
Particle density (g cm^{-3})	2.63	2.63	NS	+0.26
Water-holding capacity (%)	11.73	9.84	NS	+19.20
Porosity (%)	34.64	33.64	NS	+2.97

Table 10. Comparison of physical parameters of soil under organic vs conventional management in taro

Chemical parameters	EFY				Yams			
	Organic	Conventional	CD (0.05)	% increase or decrease	Organic	Conventional	CD (0.05)	% increase or decrease
pH	5.32	4.55	0.285	+0.77 unit	5.47	5.01	0.212	+0.46 unit
Organic C (%)	1.40	1.18	NS	+19.02	0.86	0.75	NS	+14.00
Available N (kg ha^{-1})	125.60	103.30	NS	+21.59	193.00	162.00	NS	+19.14
Available P (kg ha^{-1})	65.20	57.30	NS	+13.13	270.00	289.00	NS	−6.57
Available K (kg ha^{-1})	362.00	340.90	NS	+6.19	343.50	256.40	40.21	+33.97

Table 11. Comparison of chemical parameters of soil under organic vs conventional management in EFY and yams

The population of bacteria was considerably higher in organic plots than in conventional plots; 41 and 23% higher in EFY and yams, respectively. Organic farming also favoured the fungal population by 17–20%. While the N fixers showed an upper hand in organically managed soils by 10% over conventional management under EFY, P solubilizers remained more conspicuous under organic management of yams (22% higher than conventional management) (Table 13). The dehydrogenase enzyme activity was higher by 23 and 14% in organic plots in EFY and yams (Table 13).

In these studies, the organic resources used to replace chemical fertilizers were FYM, green manure, neem cake and ash. Green manuring with cowpea (incorporation of 15–20 t ha^{-1} of green matter) was the most cost-effective component among these. The decomposition of these organic manures to release available plant nutrients involves intense microbial activity over chemical fertilizer-applied conventional plots. This might have resulted in higher microbial population and dehydrogenase enzyme activity in the organic plots. Several earlier workers also noticed increased microbial population in cultivated organically managed soil [3, 9, 60].

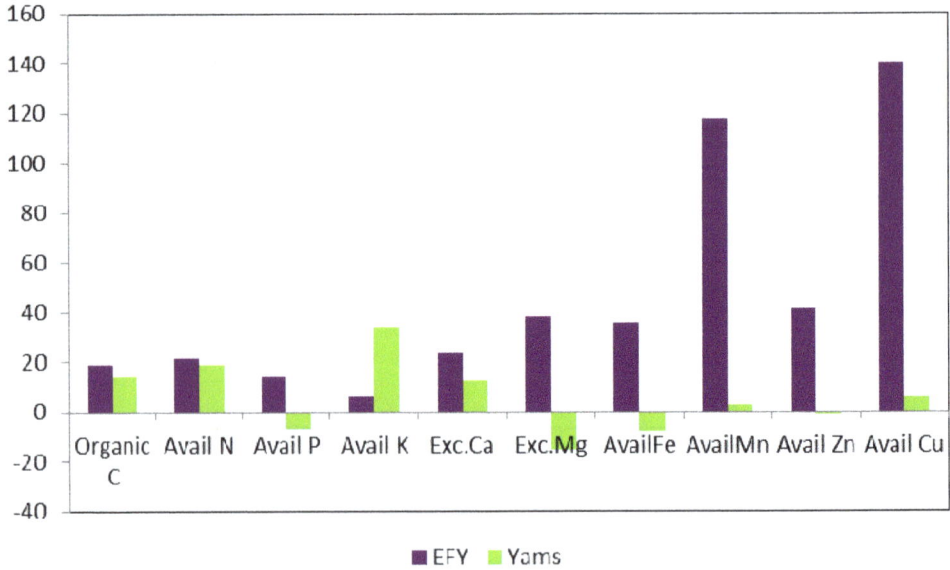

Figure 13. Per cent increase or decrease in chemical properties of soil under organic management in EFY and yams

Chemical parameters	Dwarf white yam				Taro			
	Organic	Conventional	CD (0.05)	% increase or decrease	Organic	Conventional	CD (0.05)	% increase or decrease
pH	5.68	4.56	0.467	+1.11 unit	6.68	5.48	0.473	+1.20 unit
Organic C (%)	2.29	1.97	NS	+16.46	0.84	0.60	NS	+39.03
Available N (kg ha^{-1})	119.80	109.80	NS	+9.11	105.00	103.50	NS	+1.45
Available P (kg ha^{-1})	107.30	98.00	NS	+9.49	75.10	47.10	20.34	+59.44
Available K (kg ha^{-1})	453.00	312.00	NS	+45.19	148.00	202.00	NS	−26.73

Table 12. Comparison of chemical parameters of soil under organic vs conventional management in dwarf white yam and taro

8.2.4.2. Development of SQI

In EFY, the organic system scored a significantly higher SQI (1.930), closely followed by the traditional system (1.913) (Figure 14). The SQI of conventional (1.456) and biofertilizer systems (1.580) were significantly lower. The SQI was driven by water-holding capacity, pH and available Zn followed by SOM.

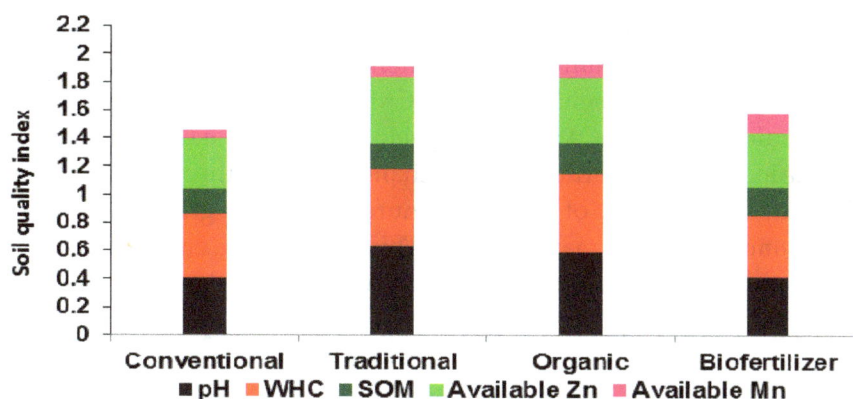

Figure 14. Effect of production systems on SQI in EFY (Source: Reference [13])

Biological Parameters	EFY				Yams			
	Organic	Conventional	CD (0.05)	Per cent increase (+) or decrease (-) in organic farming	Organic	Conventional	CD (0.05)	Per cent increase (+) or decrease (-) in organic farming
Bacteria (cfu g^{-1} soil)	31×10^7	22×10^7	NS	+40.90	118×10^3	96×10^3	NS	+22.91
Fungi (cfu g^{-1} soil)	6×10^6	5×10^6	NS	+20.00	7×10^2	6×10^2	NS	+16.66
Actinomycetes (cfu g^{-1} soil)	22×10^5	24×10^5	NS	−8.33	11×10^3	12×10^3	4.682	−8.33
N fixers (cfu g^{-1} soil)	182×10^5	165×10^5	NS	+10.30	7×10^3	11×10^3	NS	−36.36
P solubilizers (cfu g^{-1} soil)	5×10^6	5×10^6	NS	0	11×10^3	9×10^3	NS	+22.22
Dehydrogenase enzyme (μg TPF formed g^{-1} soil h^{-1})	1.625	1.323	NS	+22.82	1.174	0.786	NS	+49.36

Source: Reference [14]

Table 13. Comparison of biological parameters of soil under organic vs conventional management in EFY and yams

Soil quality is the capacity of a soil to function within natural or managed ecosystem boundaries to sustain plant and animal productivity in order to maintain or enhance water and air quality and support human health and habitation [61]. In this study, organic farming, which is a supplemental C management practice (SCMP) significantly changed a number of soil properties including soil pH, SOM, exchangeable Mg, available Cu, Mn and Fe contents and WHC. Thus, the indicator properties could be changed mainly through SOM building practices brought about by the strict use of organic manures especially green manuring continuously for five years under organic management. This framework emphasizes that soil quality assessment is a tool that can be used to evaluate the effects of land management on soil function.

9. On-farm validation of organic production technologies

Demonstration trials were conducted during 2008–2009 in 10 farmers' sites to cover an area of 5 ha in Kollam and Pathanamthitta districts of Kerala to compare the yield, quality, economics and soil fertility under the organic management practices with the existing farmers' practice and conventional practice (present package of practices recommendations) in EFY (Figure 15). Organic farming resulted in higher corm yield (34.60 t ha^{-1}) and additional income (Rs. 43,651 ha^{-1}) over conventional farming. Organically produced corms had significantly higher dry matter and Mg contents and significantly lower oxalate content. The chemical properties of the soil, especially K, was seen to be favoured under organic farming (Table 14).

Production systems	Yield (t ha^{-1})	Corm dry matter (%)	Oxalate (DW basis %) content of corms	Mg content of corms (mg 100 g)	Available K of soil (kg ha $^{-1}$)	Net income (Rs ha^{-1})	B:C ratio
Conventional	24.50	19.29	0.221	91.90	98.80	70,069	1.40
Traditional	22.20	20.00	0.218	91.80	88.70	41,925	1.23
Organic	34.60	21.00	0.191	95.30	142.70	1,13,720	1.49
CD (0.05)	7.750	1.162	0.0076	2.045	40.02		

Source: Reference [11]

Table 14. Agronomic, nutritional and economic implications of organic management in EFY under validation trials

OFT were laid out in seven sites with three practices, conventional, traditional and organic, in Thiruvananthapuram and Kollam districts of Kerala to validate the on-station-developed organic farming technologies in yams (greater yam, lesser yam and dwarf white yam) and taro (Figure 15). In all sites, tuber yield under organic management was on a par with conventional practice in these crops (Figure 16). However, the yields under organic management were 8, 17, 21 and 29% higher over chemical-based farming in greater yam, lesser yam, dwarf white yam and taro, respectively. In general, there was significant improvement in pH, organic C and available K status under organic management in the sites. Soil microbial population was also improved under organic practice in these sites.

Farmers convinced about green manuring

Organic corms of EFY

View of OFT on yams and taro

Farmers with organic tubers of greater yam

Figure 15. On-farm validation trials conducted in Kerala

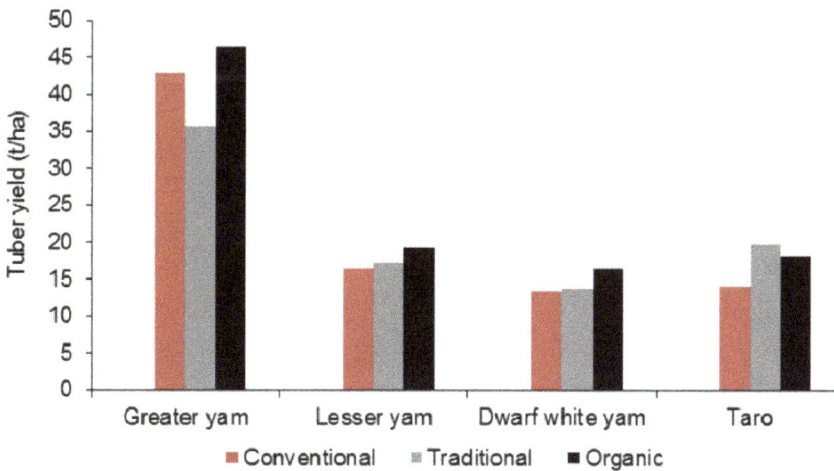

Figure 16. Yield under various practices in OFT in yams and taro

10. The package

Use of organically produced seed materials, seed treatment in cow-dung, neem cake, bio-inoculant slurry, FYM incubated with bio-inoculants, green manuring, use of neem cake, bio-fertilizers and ash formed the strategies for organic production (Figure 17). The organic farming package for EFY is included in the Package of Practices Recommendations for crops by Kerala Agricultural University [62].

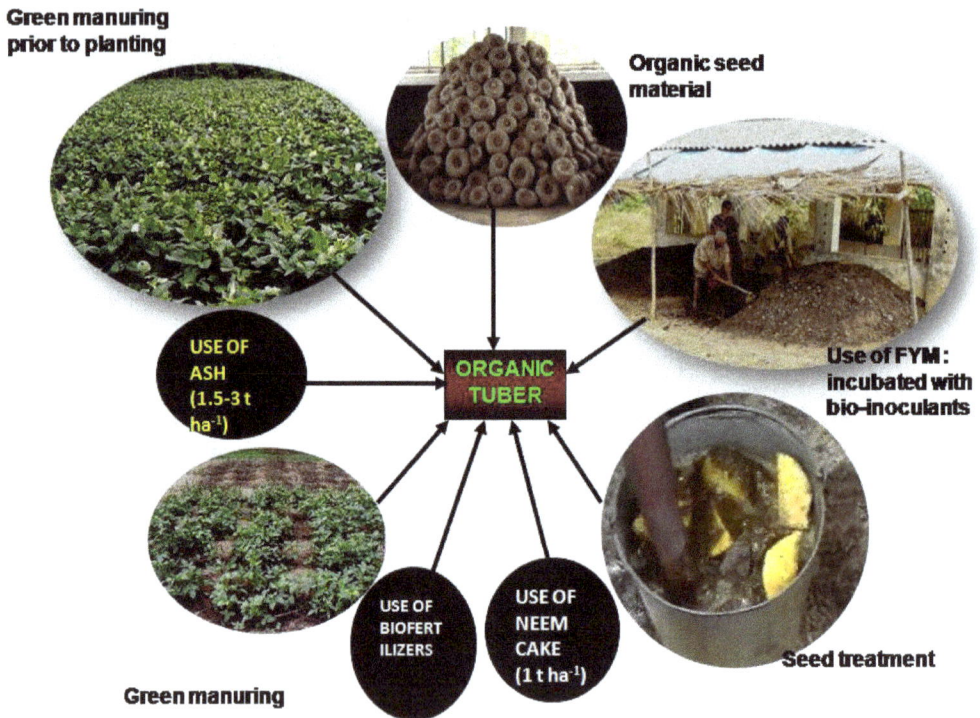

Figure 17. Essential components of organic tuber production

11. Development of a learning system

A learning system was developed using ANN to predict the performance of EFY production system. A three-layered system with one input layer, one output layer and one hidden layer was developed. The input layer neurons included temperature, rainfall, planting material, FYM, potassium, phosphorus, ash, neem cake, *Azospirillum*, phosphobacteria, mycorrhiza and green manure. The output layer neurons were total biomass, corm yield, canopy spread and plant height.

11.1. Structure of the system

A three-layered feed-forward back-propagation network (FFBPN) (Figure 18) was designed for this learning system [41]. Its block diagram (Figure 19) explains the flow of the inputs and the modifications made on it while it passes through the different layers before the output is generated.

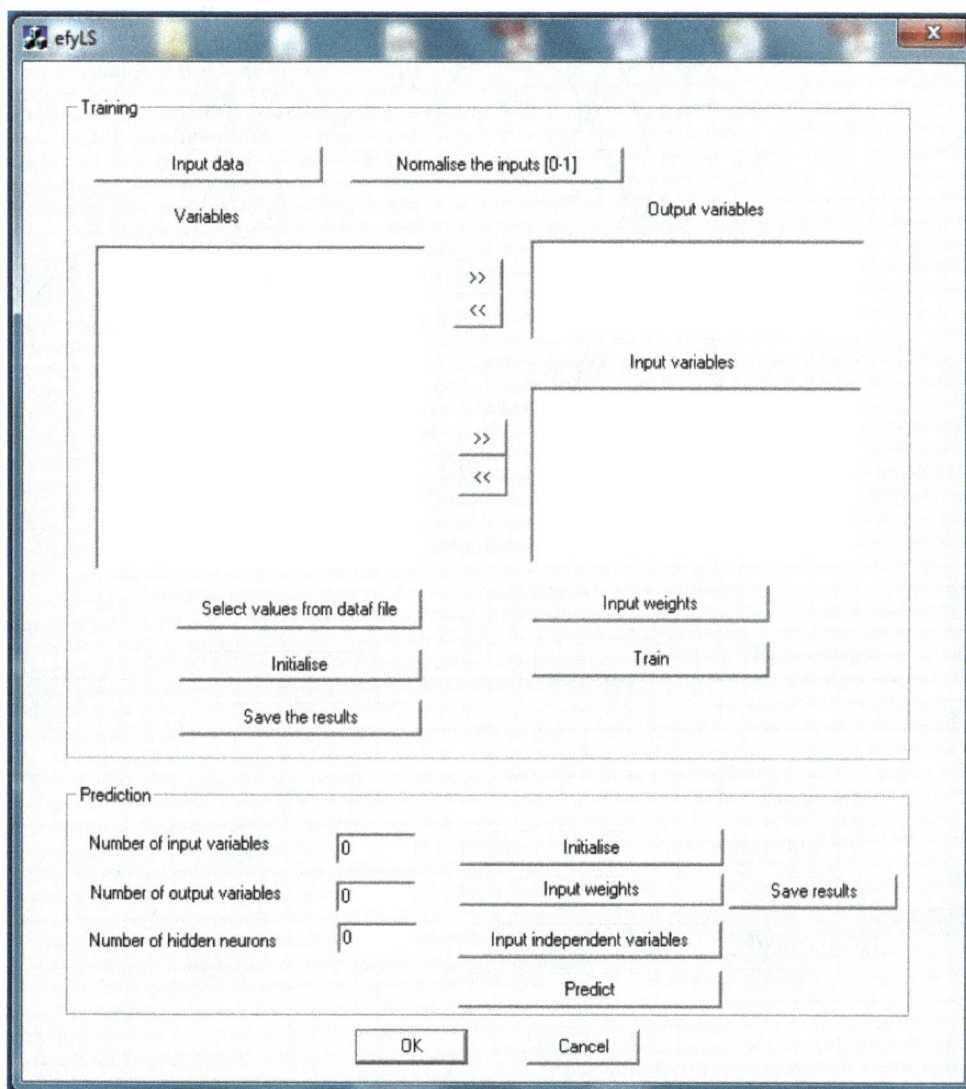

Figure 18. Learning system to predict the performance of EFY production system

Input layer of the network is composed of 12 neurons represented by I1, I2,..., I12. The activities of neurons in the input layer represent the raw information that is fed into the network. Inputs added to the neurons of the input layer are given in Table 15.

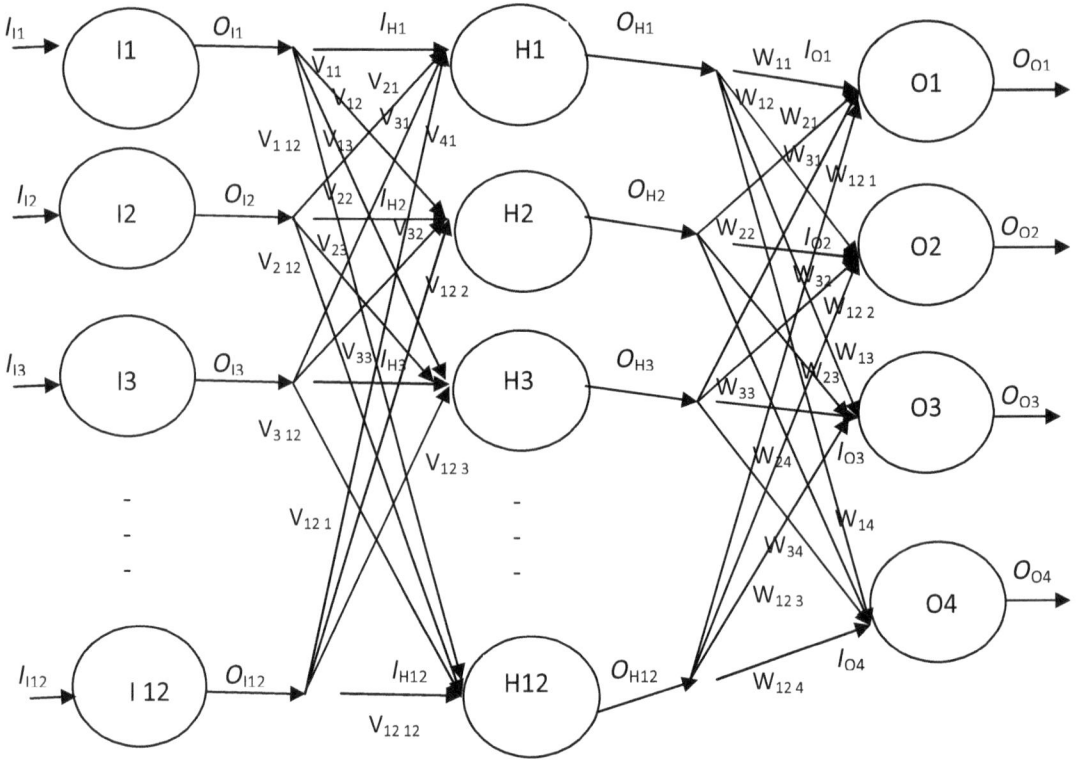

Figure 19. Structure of the three-layered FFBPN of the learning system

Sl. No.	Inputs added	Neuron of the input layer
1.	Temperature (°C)	I1
2.	Rainfall (mm)	I2
3.	Planting material (kg)	I3
4.	Farmyard manure (kg)	I4
5.	Potassium (kg)	I5
6.	Phosphorus (kg)	I6
7.	Ash (kg)	I7
8.	Neem cake (kg)	I8
9.	*Azospirillum* (kg)	I9
10.	Phosphobacteria (kg)	I10
11.	Mycorrhiza (kg)	I11
12.	Green manure (kg)	I12

Table 15. List of inputs added to various neurons in the input layer of the FFBPN

As linear activation function is operating in the input layer of the network, the input (I) and output (O) of the input layer are the same:

$$\{O\}_t = \{I\}_t \tag{1}$$

The hidden neurons H1...H12 are connected by synapse to the input neurons. Let $V_{m,p}$ be the weight of the arc between mth input neuron and the pth hidden neuron. The input to the hidden neuron is the weighted sum of the outputs of the input neurons to get I_{Hp}, i.e. the input to the pth hidden neuron as

$$I_{Hp} = \sum_{m=1,p=1}^{12,12} V_{m,p} O_{I,m} \tag{2}$$

where

$O_{I,m}$ is the output of mth input neuron.

In the hidden neurons, sigmoidal function is operating and thus the output of the pth hidden neuron is given by

$$O_{Hp} = \frac{1}{(1 + e^{-\lambda(I_{Hp} - \theta_{Hp})})} \tag{3}$$

where

O_{Hp} is the output of the pth hidden neuron

I_{Hp} is the input of the pth hidden neuron and

θ_{Hp} is the threshold of the pth hidden neuron, which is initialized to zero in this system

Input to the output neurons is the weighted sum of the outputs of the hidden neurons. Input to the qth output neuron I_{Oq} is calculated as follows:

$$I_{Oq} = \sum_{n=1,q=1}^{12,4} W_{n,q} O_{H,n} \tag{4}$$

where

O_{Hn} is the output of the nth hidden neuron and

$W_{n,q}$ is the weight of the arc between nth hidden neuron and qth output neuron.

Sigmoidal function is operating in the output neurons also, and the output of the qth neuron is given by

$$O_{Oq} = \frac{1}{(1 + e^{-\lambda(I_{Oq} - \theta_{Oq})})} \tag{5}$$

where

O_{Oq} is the output of the qth output neuron

I_{Oq} is the input of the qth output neuron and

θ_{Oq} is the threshold of the qth output neuron which is initialized to zero in this system

11.2. Training of the system

A three-layered FFBPN was designed for this learning system. Three years data (Table 16) on various aspects of cultivation of EFY were used for training the system.

Inputs	Input neuron	Years		
		2004	2005	2006
Temperature (°C)	I1	27.91	28.68	27.91
Rainfall (mm/season)	I2	2,179.90	1,862.95	2,082.45
Planting material (kg)	I3	750	750	750
FYM (t ha^{-1})	I4	25	25	25
Nitrogen (kg ha^{-1})	I5	100	100	100
Phosphorus (kg ha^{-1})	I6	50	50	50
Potassium (kg ha^{-1})	I7	150	150	150
Ash (kg ha^{-1})	I8	0	0	0
Neem cake (kg ha^{-1})	I9	0	0	0
Azospirillum (kg ha^{-1})	I10	0	0	0
Phosphobacteria (kg ha^{-1})	I11	0	0	0
Mycorrhiza (kg ha^{-1})	I12	0	0	0
Outputs	**Output neuron**	**2004**	**2005**	**2006**
Total biomass (kg plant^{-1})	O1	3.48	3.19	3.14
Corm yield (kg plant^{-1})	O2	2.95	2.83	2.93
Canopy spread (cm plant^{-1})	O3	110.37	111.61	101.18
Plant height (cm plant^{-1})	O4	55.87	62.54	48.72

Table 16. Values used for training the learning system

Weight matrix obtained between input and hidden layers and between hidden and output layers is stored in the database and is used for making predictions with other input data-sets. This system learns about the EFY production system when the independent variables like weather parameters, soil and nutritional parameters of the system as well as the corresponding dependent variables of the system like com yield, canopy size, etc., are fed as input into it. Once it learns about a particular system pattern, it can predict the outputs corresponding to another set of independent variables of a similar pattern. The system can be trained for various independent–dependent variable patterns so that dependent variables for another set of same independent variables can be predicted accurately. When more and more inputs are used for training as well as prediction, the system learns more and its precision increases.

12. Constraints in promotion of organic farming

In India, the availability of organic manures is a major constraint. It is estimated that to feed 1.4 billion population by the year 2025, a minimum of 301 million tonnes of food grains are needed. To meet this demand, it will be necessary to harness 30–35 million tonnes of NPK from fertilizer carriers and an additional 10 million tonnes from organic and biofertilizer sources [63]. Thus, only approximately 25–30% nutrient needs of Indian Agriculture can be met by utilizing organic sources solely [24, 64]. Organic manures are bulky (high cost of handling and transportation), of low analysis, slowly available and variable in composition. The availability of cattle dung for organic farming will be further limited as this is a major source of fuel in rural households. Apart from these, green manuring and recycling of farm wastes as manures have not become popular as these are more time and space consuming and their impacts on productivity are not rapidly discernible. At present, certification procedures are cumbersome and expensive [24, 64].

13. Future thrust

Some of the future lines of action for promotion of organic farming have been identified [24, 64, 65]. Proper delineation and identification of prospective areas and crops (like tuber crops) may be helpful for effective promotion of organic farming. There is a need to undertake systematic research on the comparative values/advantages of organic farming over conventional farming on a long-term basis for promotion of organic farming. The package of practices recommendations for organic farming has to be popularized. The extent of availability of potential organic sources needs to be ascertained along with measures that may be helpful in improving the convenience of their use. Environmental impact, especially water and air quality effects, of organic farming needs to be assessed.

Weed management options particularly under climate change by nonchemical and biological methods are limited and need evaluation. The benefits accruing through organic farming on crop yield, quality, market preference and price advantage may be properly understood and promoted among the farmers and consumers [24].

14. Conclusions

In order to attain sustainable food-cum-livelihood-cum-environmental security in India, we may require an array of alternatives to chemical intensive agriculture. Instead of seriously debating on organic vs conventional agriculture it is better to examine critically the costs and benefits of the different alternative management options. It has been conclusively proved in tuber crops that organic management is an alternative viable option for sustainable and safe food production with less soil degradation and environmental pollution. Tuber crops, especially EFY and yams are prospective candidates for organic farming. EFY is the most responsive, followed by greater yam, white yam, lesser yam and taro. Generation of sufficient biomass, addition of crop residues, green manuring, farm waste recycling, fortification of manures through proper composting, adoption of crop rotations involving legumes, establishment of biogas plants and development of agro-forestry for alternate source of fuels are some of the strategies that will help promote organic farming of tuber crops. These practices would help a great deal in supplementing/rationalizing the use of inorganic fertilizers, which cannot be totally eliminated in Indian Agriculture.

Author details

Suja Girija, Sreekumar Janardanan, Jyothi Alummoottil Narayanan and
Santosh Mithra Velayudhan Santhakumari

*Address all correspondence to: sujagin@yahoo.com

ICAR-Central Tuber Crops Research Institute, Thiruvananthapuram, Kerala, India

References

[1] Carter ME, Gamez R, Gliessman S. Sustainable Agriculture and the Environment in the Humid Tropics. Washington, DC: National Academy Press; 1993. 191 p.

[2] The Hindu. Climate change could trigger global food crisis. In: The Hindu. 1 September 2007. Op-Ed-pp.11.

[3] Suja G, Sreekumar J. Implications of organic management on yield, tuber quality and soil health in yams in the humid tropics. International Journal of Plant Production. 2014; 8(3): 291–309.

[4] Susan Andrews S, Mitchell JP, Mancinelli R, Karlen DL, Hartz TK, Horwath WR, Pettygrove GS, Scow KM, Munk DS. On-farm assessment of soil quality in California's central valley. Agronomy Journal. 2002; 94: 12–23.

[5] Chhonkar PK. Organic farming and its relevance in India. In: Tarafdar JC, Tripathi KP, Mahesh K, editors. Organic Agriculture. Jodhpur: Scientific Publishers; 2008. pp. 5–33.

[6] Katsvairo TW, Wright DL, Marois JJ, Rich JR. Transition from conventional farming to organic farming using bahiagrass. Journal of the Science of Food and Agriculture. 2007; 87: 2751–2756.

[7] Rembialkowska E. Quality of plant products from organic agriculture. Journal of the Science of Food and Agriculture. 2007; 87: 2757–2762.

[8] Reganold JP, Glover JDP, Andrews K, Hinman HR. Sustainability of three apple production systems. Nature. 2001; 410: 926–929.

[9] Stockdale EA, Lampkin NH, Hovi M, Keating R, Lennartsson EKM, Macdonald DW, Padel S, Tattersall FH, Wolfe MS, Watson CA. Agronomic and environmental implications of organic farming systems. Advances in Agronomy. 2001; 70: 261–327.

[10] Suja G, Susan John K, Sundaresan S. Potential of tannia (*Xanthosoma sagittifolium* L.) for organic production. Journal of Root Crops. 2009; 35(1): 36–40.

[11] Suja G, Susan John K, Ravindran CS, Prathapan K, Sundaresan S. On farm validation of organic farming technology in elephant foot yam (*Amorphophallus paeoniifolius* Dennst. Nicolson). Journal of Root Crops. 2010; 36: 59–64.

[12] Suja G, Sundaresan S, Susan John K, Sreekumar J, Misra RS. Higher yield, profit and soil quality from organic farming of elephant foot yam. Agronomy for Sustainable Development. 2012a; 32: 755–764. DOI:10.1007/s13593-011-0058-5.

[13] Suja G, Sreekumar J, Susan John K, Sundaresan S. Organic production of tuberous vegetables: Agronomic, nutritional and economic benefits. Journal of Root Crops. 2012b; 38(2): 135–141.

[14] Suja G. Comparison of tuber yield, nutritional quality and soil health under organic versus conventional production in tuberous vegetables. Indian Journal of Agricultural Sciences. 2013; 83(11): 35–40.

[15] Branca G, Lipper L, McCarthy N, Jolejole MC. Food security, climate change, and sustainable land management. A review. Agronomy for Sustainable Development. 2013; 33: 635–650.

[16] UNFPA. The state of the world population report. By choice, not by chance: family planning, human rights and development. New York, NY: United Nations Population Fund; 2012.

[17] IAASTD. Agriculture at the crossroads. International Assessment of Agricultural Knowledge, Science and Technology for Development. Washington, DC: Island Press; 2009.

[18] Tilman D, Cassman KG, Matson PA, Naylor R, Polasky S. Agricultural sustainability and intensive production practices. Nature. 2002; 418: 671–677. DOI:10.1038/nature01014.

[19] Lal R. Degradation and resilience of soils. Philosophical Transactions of the Royal Society of London B. 1997; 352: 869–889. DOI:10.1098/rstb.1997.0078.

[20] World Bank. Sustainable land management: challenges, opportunities, and trade-offs. Washington, DC: The World Bank; 2006.

[21] Woodfine A. The potential of sustainable land management practices for climate change mitigation and adaptation in sub-Saharan Africa. Rome: Food and Agriculture Organization of the United Nations; 2009.

[22] Pretty JN. Agricultural sustainability: concepts, principles and evidence. Philosophical Transactions of the Royal Society of London B. 2008; 363(1491): 447–466. DOI: 10.1098/rstb.2007.2163.

[23] Pretty JN. Editorial: sustainable intensification in Africa. International Journal of Agricultural Sustainability. 2011; 9(1): 3–4. DOI:10.3763/ijas.2010.0583.

[24] Suja G. Strategies for organic production of tropical tuber crops. In: Venkateswarlu B, Balloli SS, Ramakrishna YS, editors. Organic Farming in Rainfed Agriculture: Opportunities and Constraints. Hyderabad: Central Research Institute for Dryland Agriculture; 2008. pp. 135–143

[25] Das S, Biswas BC. Organic farming-prospects and problems. Fertilizer News. 2002; 47(12): 105–112, 115–118.

[26] Susan John K, Mohankumar CR, Ravindran CS, Prabhakar M. Long term effects of manures and fertilizers on cassava production and soil productivity in an acid ultisol. In: Proceedings National Workshop on Long Term Soil Fertility Management through Integrated Plant Nutrient Supply; 21–25 June 1998; Bhopal, India; 1998. pp. 318–325.

[27] Ravindran CS, Bala N. Effect of FYM and NPK on the yield and quality of sweet potato. Journal of Root Crops. 1987; 13(1): 35–39.

[28] Suja G. 2001. Resource management for intercropping white yam (*Dioscorea rotundata* Poir.) in coconut garden. [Ph.D. thesis]. Thrissur: Kerala Agricultural University; 2001. 195 p.

[29] Nayar TVR, Suja G. Organic production potential of tropical tuber crops. In: Abstracts XIV Swadeshi Science Congress; 5–7 November 2004; Kochi: Swadeshi Science Movement; 2004. pp. 20–21.

[30] AOAC. Official Methods of Analysis. Washington, DC: Association of Official Agricultural Chemists; 1980. pp. 169, 611–612, 1141.

[31] Dubois M, Gilles KA, Hamilton JK, Rebers PA, Smith F. Colorimetric method for determination of sugars and related substances. Analytical Chemistry. 1956; 28: 350–356.

[32] Nelson N. A photometric adaptation of the Somogyi method for determination of glucose. Journal of Biology and Chemistry.1944; 153: 375–380.

[33] Swain T, Hillis WE. The phenolic constituents of *Prumus domestica* 1. The quantitative analysis of phenolic constituents. Journal of the Science of Food and Agriculture. 1955; 10: 963–968.

[34] Piper, CS. Plant and Soil Analysis. Bombay: Hans Publications; 1970.

[35] Page AL, Miller RH, Keeney DR. Methods of Soil Analysis, Part 2: Chemical and Microbiological Properties, Agronomy Series No. 9, Madison, WI: American Society of Agronomy, Soil Science Society of America; 1982. 1159 p.

[36] Gupta RP, Dakshinamoorthy C. Procedures for Physical Analysis of Soil and Collection of Agrometeorological Data. New Delhi: Indian Agricultural Research Institute; 1980.

[37] Timonin MI. The interaction of higher plants and soil microorganisms I. Microbial population of the rhizosphere of seedlings of certain cultivated plants. Canadian Journal of Research. 1940; 181: 307–317.

[38] Klein DA, Loh TC, Goulding RL. A rapid procedure to evaluate dehydrogenase activity of soils low in organic matter. Soil Biology and Biochemistry. 1971; 3: 385–387.

[39] Karlen DL, Stott DE. A framework for evaluating physical and chemical indicators of soil quality. In: Doran JW, Coleman DC, Bezdicek DF, Stewart BA, editors. Defining Soil Quality for a Sustainable Environment. SSSA Special Publication No. 35. Madison, WI: Soil Science Society of America; 1994. pp. 53–72.

[40] SAS. SAS Users Guide. Cary, NC: SAS Institute Inc.; 2008.

[41] Frank Z, David B. Plant growth model using artificial neural networks. 1997. http://trs new.jpl.nasa.gov/dspace/bitstream/2014/22170/1/97-0632.pdf

[42] Rajasekaran S, Vijayalakshmi Pai GA. Neural Networks, Fuzzy Logic and Genetic Algorithms Synthesis and Applications. Delhi: PHI Publications; 2009.

[43] Ponti TD, Rijk B, van Ittersum MK. The crop yield gap between organic and conventional agriculture. Agricultural Systems. 2012; 108: 1–9.

[44] Seufert V, Ramankutty N, Foley JA. Comparing the yields of organic and conventional agriculture. Nature. 2012; 485: 229–232.

[45] Ramesh P, Mohan S, Subha Rao A. Organic farming: its relevance to the Indian context. Current Science. 2005; 88: 561–568.

[46] Ramesh P, Panwar NR, Singh AB, Ramana S, Yadav SK, Shrivastava R, Subha Rao A. Status of organic farming in India. Current Science. 2010; 98: 1190–1194.

[47] Suja G, Santhosh Mithra VS, Sreekumar J, Jyothi AN. Is organic tuber production promising? Focus on implications, technologies and learning system development. In: Rahmann G, Aksoy U, editors. Proceedings of the 4th ISOFAR Scientific Conference. 'Building Organic Bridges'. Organic World Congress; 13–15 October 2014; Istanbul, Turkey; 2014. eprint ID 23666.

[48] Stanhill G. The comparative productivity of organic agriculture. Agriculture Ecosystem and Environment. 1990; 30: 1–26.

[49] Offermann F, Nieberg H. Economic performance of organic farms in Europe. In: Organic Farming in Europe: Economics and Policy, vol. 5. Hohenheim: University of Hohenheim; 1999.

[50] Clark MS, Horwath WR, Shennan C, Scow KM. Changes in soil chemical properties resulting from organic and low-input farming systems. Agronomy Journal. 1998; 90: 662–671.

[51] Colla G, Mitchell JP, Joyce BA, Huyck LM, Wallender W, Temple SR, Hsiao TC, Poudel DD. Soil physical properties and tomato yield and quality in alternative cropping systems. Agronomy Journal. 2000; 92: 924–932.

[52] Kabeerathumma S, Mohan Kumar B, Nair PG. Nutrient uptake and their utilization by yams, aroids and coleus. Technical Bulletin Series No. 10. Thiruvananthapuram: Central Tuber Crops Research Institute; 1987.

[53] Pieper JR, Barrett DM. Effects of organic and conventional production systems on quality and nutritional parameters of processing tomatoes. Journal of the Science of Food and Agriculture. 2008; 89: 177–194.

[54] Worthington V. Nutritional quality of organic versus conventional fruits, vegetables and grains. The Journal of Alternative and Complementary Medicine. 2001; 7(2): 161–173.

[55] Droogers P, Fermont A, Bouma J. Effects of ecological soil management on workability and trafficability of a loamy soil in the Netherlands. Geoderma. 1996; 73: 131–145.

[56] Gerhardt RA. A comparative analysis of the effects of organic and conventional farming systems on soil structure. Biological Agriculture Horticulture. 1997; 14: 139–157.

[57] Mei C, Yahua S, Zhengguo O, Qirong S. Amelioration of aluminium toxicity with pig manure in an acid red soil. Plant Nutrition and Fertilizer Science. 2002; 8(2): 176.

[58] Prabhakaran C, Pitchai GJ. Effect of different organic nitrogen sources on pH, total soluble solids, titrable acidity, reducing and non reducing sugars, crude protein and

ascorbic acid content of tomato fruits. Journal of Soils and Crops. 2002; 12(2): 160–166.

[59] Prakash YS, Bhadoria PBS, Rakshit A. Comparative efficacy of organic manures on the changes in soil properties and nutrient availability in an Alfisol. Journal of the Indian Society of Soil Science. 2002; 50(2): 219–221.

[60] Scow KM, Somasco O, Gunapala N, Lau S, Venette R, Ferris H, Miller R, Shenman C. Transition from conventional to low-input agriculture changes soil fertility and biology. California Agriculture. 1994; 48: 20–26.

[61] Karlen DL, Mausbach MJ, Doran JW, Cline RG, Harris RF, Sehuman GE. Soil quality: A concept, definition and framework for evaluation. Soil Science Society of America Journal. 1997; 61: 4–10.

[62] KAU. Package of Practices Recommendations: Crops. Thrissur: Kerala Agricultural University; 2011. pp. 49–50.

[63] NAAS. Fertilizer Policy Issues (2000–2025). Policy Paper 2. New Delhi: National Academy of Agricultural Sciences; 1997. 5 p.

[64] Chhonkar PK, Dwivedi BS. Organic farming and its implications on India's food security. Fertilizer News. 2004; 49(11): 15–18, 21–28, 31, 38.

[65] Marwaha BC, Jat SL. Status and scope of organic farming in India. Fertilizer News. 2004; 49(11): 41–48.

7

Potatoes (*Solanum tuberosum* L.)

Petr Dvořák, Jaroslav Tomášek, Karel Hamouz and Michaela Jedličková

Additional information is available at the end of the chapter

Abstract

In the area of potato production, targeted research solving concrete and actual problems of potato producers runs on Czech University of Life Sciences in Prague. In the last few years, we were focused on the production of new potatoes designated for early harvest, and we were focused on capitalization of yielding and qualitative characteristics of colored potato variety. These findings were further utilized and transferred to the system of organic farming. Firstly, we watched the influence of organic farming on yield and quality of tubers. Ecological ways of cultivation had strong negative influences on yield (decrease of 36%). From qualitative characteristics, organic farming increased the content of polyphenols by 10.2%, decreased the content of nitrates by 11.0%, and decreased the content of reducing sugars by 22.0%. We also evaluated the possibilities and impacts of mulch on potato cultivation. The mulch on top of ridges affected the temperature of soil (it increased the temperature by 0.2–0.6 °C under black mulching nonwoven fabric, and it decreased by 0.5–0.8 °C under herbal mulch). The mulch also affected soil humidity (herbal mulch decreased the soil humidity) and adjust weed infestation (20 to 92% lower), soil erosion (95% lower), the occurrence of Colorado beetle (the number of larvae was 22.8% lower with herbal mulch and 88.7% higher with mulching textile), and late blight in potato vegetation.

Keywords: Potatoes, organic farming, mulching, plant extracts, quality tubers

1. Introduction

Potatoes in Czech Republic belong to the minority crops cultivated in the system of organic farming. Like the principal tuber crop, it forms ca. 0.5% of the whole certified area of Czech Republic. The area of consumption potatoes actuate over is an area of 200 ha (in 2012, 3,277 tonnes of organic potatoes were harvested on an area of 230 ha).

Cultivating potatoes organically is very demanding on producers. Producers must deal with the absence of chemicals used on crop protection, the absence of synthetic fertilizers, the obtainment of acceptable yield and good quality of tubers, and the necessity of applying all the procedures to create suitable conditions for growth and development of crops, like any other crop cultivated organically [1].

2. Environmental conditions

The potatoes originate from the mountain area that is why the foothill conditions suit them well. The optimum amount of precipitations for potatoes is 650 to 800 mm annually (60–70% of this amount during the vegetation). The precipitations during the first half of vegetation influence the growth of tops, the precipitations from May to half of July influence the number of tubers under the clump (with consideration of the time of planting and earliness of the variety). The precipitations in the second half of vegetation determine the weight of tubers. The deficiency of precipitations during the period of planting until emergence relatively positively affects the yield of tubers. Plants produce more roots and can manage water better [1].

In case of early potatoes, where the well-timed soil preparation and well-timed planting is important (it occurs until the end of April in the Czech Republic), we choose the fields with soil easily processed early in the spring. From the point of view of the regulation of fungi diseases, we prefer the open fields (air locations which provide quick drying of plants). The good choice of location can regulate the occurrence of late blight [2].

3. Choice of suitable variety

Like any other crops, the choice of variety in the system of organic farming is crucial. The quality and health conditions of chosen planting material are vital, too. Generally recommended are the varieties with shorter vegetation period (with quicker initial growth and quicker tuber formation), lower nitrogen requirements, and higher resistance against diseases [2]. In case of varieties with longer vegetation period (usually intended for autumn consumption and storage), it is important to choose varieties highly resistant against late blight [3].

The choice of variety is submitted to the purpose of production (direct consumption, washing or peeling, on food-processing products such as chips and potato puree). For the consumer varieties, the determining aspects are qualitative indexes expressed by table value. It is commonly expressed by so-called cooking type of tubers (based on evaluation of consistency of cooked tubers, moisture, structure, mealiness, darkening, and taste). For this purpose are potato varieties divided into four groups: (1) cooking type A – consistent, tallowy, of delicate to semi-delicate structure, cannot be overcook, very weakly to weakly farinaceous tubers (suitable for preparation of potato salads or for meals when it is necessary to keep the shape even after cooking, like in case of soups, and for common consumption); (2) cooking type B – semi-consistent, semi-farinaceous, pleasantly moist to dry (suitable generally as a side dish); (3) cooking type C – soft, farinaceous tubers, semi-moist to dry (suitable mainly for preparation of purees, potato dough, and potato pancakes); (4) cooking type D – rough, strongly farina-ceous, and can be overcooked (undesirable for consumption purposes, usable for starch processing or for other products).

Until the present, no compact information is available in Czech Republic concerning the comparison of potato varieties in the system of organic farming.

The colored varieties are an interesting area for organic farming. They are more frequent in organic farms abroad. There is a speciality from the viewpoint of both appearance (colorfulness and shape of tubers) and nutritional value (mainly the high content of antioxidants and pigments). This area has been, in the long term, intensively examined by Prof. Ing. Karel Hamouz, CSc. and his colleagues from the Faculty of Agrobiology, Food and Natural Resour-ces, Czech University of Life Sciences (CULS) in Prague. Their studies are deepening the known information about these varieties (antioxidant activity, content of anthocyanin in raw and cooked tubers). It is possible to find between them perspective varieties usable for the consumption or processing (production of natural dye agents or syrups). To this group belongs variety Valfi, which originates in Czech Republic (violet variety bred in Potato Research Institute Havlíčkův Brod).

4. Innovations in cultivation techniques

4.1. Nutrition and fertilization

The need of nutrients, specifically the plant uptake, is given by the level of yield of tubers. Potatoes need, in average, 80–130 kg of Nitrogen per hectare (it is possible to count the uptake of 40–50 kg of Nitrogen, 8.8 kg of Phosphorus, 22 kg of Potassium, and 8.4 kg of Magnesium per 10 tonnes of tubers). This need is covered by applied barnyard manure, green manure, compost, cattle slurry, or digestate. Then the level of available nutrients depends on the level of biologic activity of the soil, i.e., mineralization conditions (which are supportable by hoeing). It is also possible to enhance the biological procedures in soil by many preparations on the basis of nitrogen fixators such as Azoter or AlgaSoil-natural organic fertilizer made of seaweed. These preparations were tested in small-plot experiments on CULS's land.

4.1.1. Experimental verification

Azoter was applied by spraying a dose of 10 liters per hectare to the furrows during hand-planting. AlgaSoil was applied to the furrows near tubers in a dose of 70 kg per hectare during planting. During vegetation, the content of chlorophyll was measured by hand using the Chlorophyll Meter SPAD 502 (in five terms from the 56th to the 100th day after planting), and in case of preparation, AlgaSoil leaf samples were taken twice for analyses of nitrogen and other nutrients. After harvest, tubers were sorted by size into two groups (tubers under 4 cm and over 4 cm).

The application of Azoter supported nitrogen fixation in the soil, thanks to the three genus of nonsymbiotic bacteria contained in this preparation (*Azotobacter chroococcum, Azospirillum braziliense*, and *Bacterium megatherium*). This was also shown in plants with higher chlorophyll content in their leaves (Figure 1). The application of Azoter had positive effect on the yield of tubers that was higher by 1.1 t per hectare in comparison with untreated control (Figure 2).

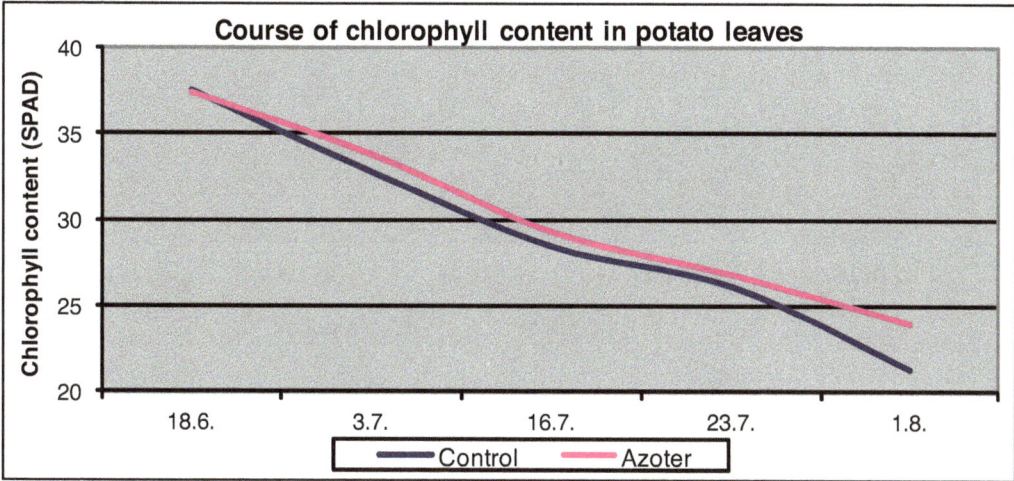

Figure 1. Chlorophyll content in potato leaves of Katka varieties in 2013 when measured by Chlorophyll Meter SPAD 502.

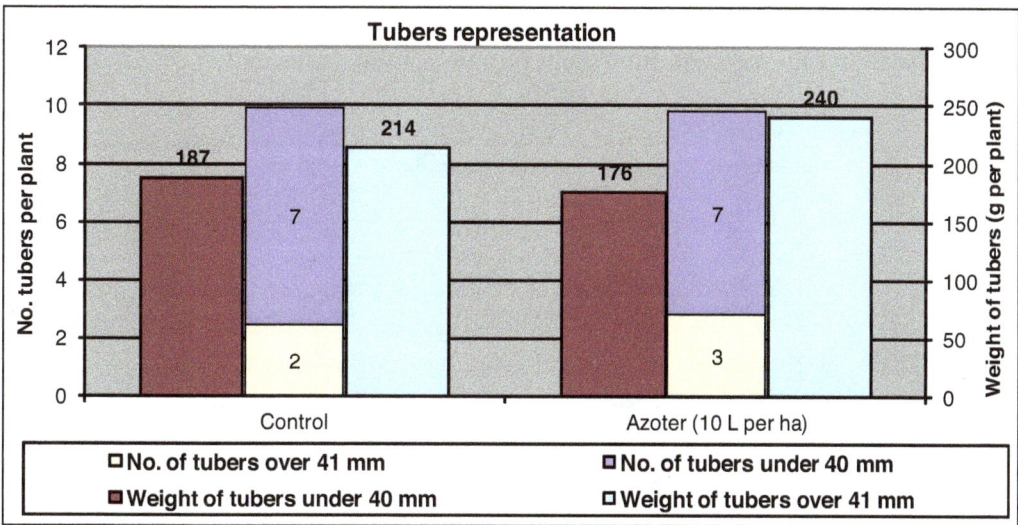

Figure 2. The final effect of Azoter on the numerical representation and weight of tubers under a clump of Katka variety in 2013.

AlgaSoil is a natural organic granulated fertilizer based on seaweed, which should work as a soil conditioner, ameliorate the soil structure, and increase the microbial activity and the utility of nutrients in soil. AlgaSoil also increased the chlorophyll content in leaves (Figure 3).

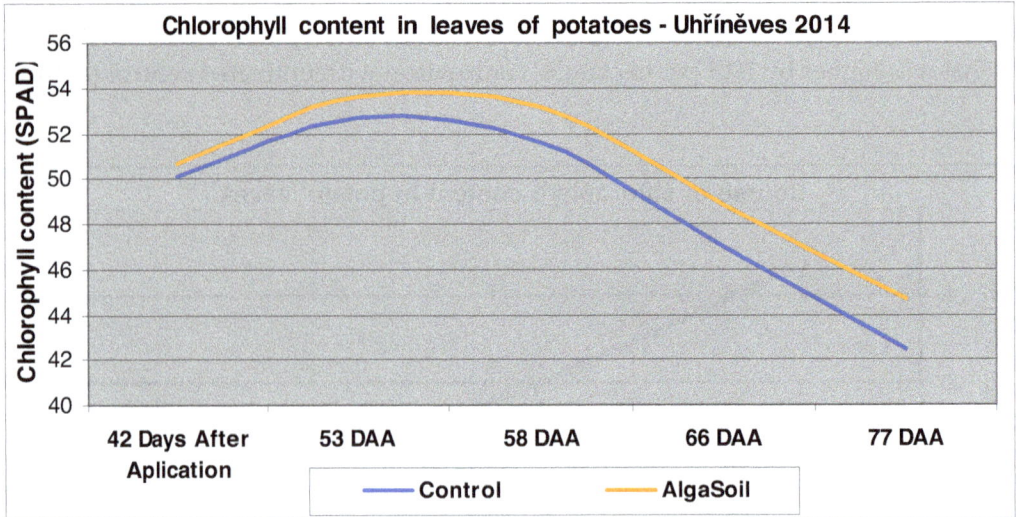

Figure 3. Course content of chlorophyll in the leaves after application of the fertilizer AlgaSoil.

There is known positive correlation between chlorophyll content and N content in plants [4]. The N content in variants treated by AlgaSoil (Figures 4 and 5) was 6% higher than in controls after first sampling (58th day after planting) and 24% higher after the second sampling (77th day after planting). Similarly, the chlorophyll content was higher on the 58th day by 3% as well as on the 77th day.

The AlgaSoil affected the size and final yield of tubers, which was higher by 3.6% (Figure 6).

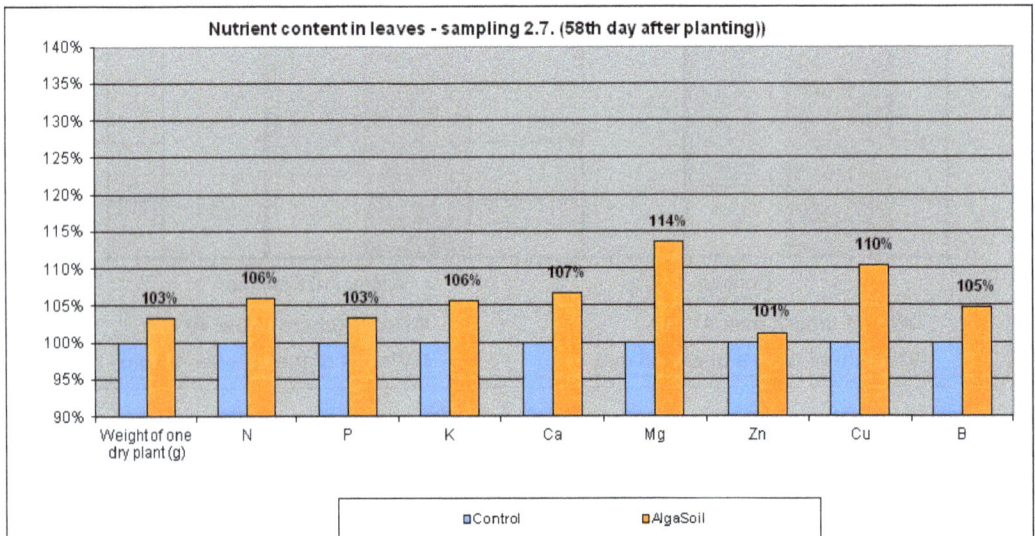

Figure 4. Results of laboratory analyses on the primary nutrients content in the leaves of potatoes on the 58th day after planting [14].

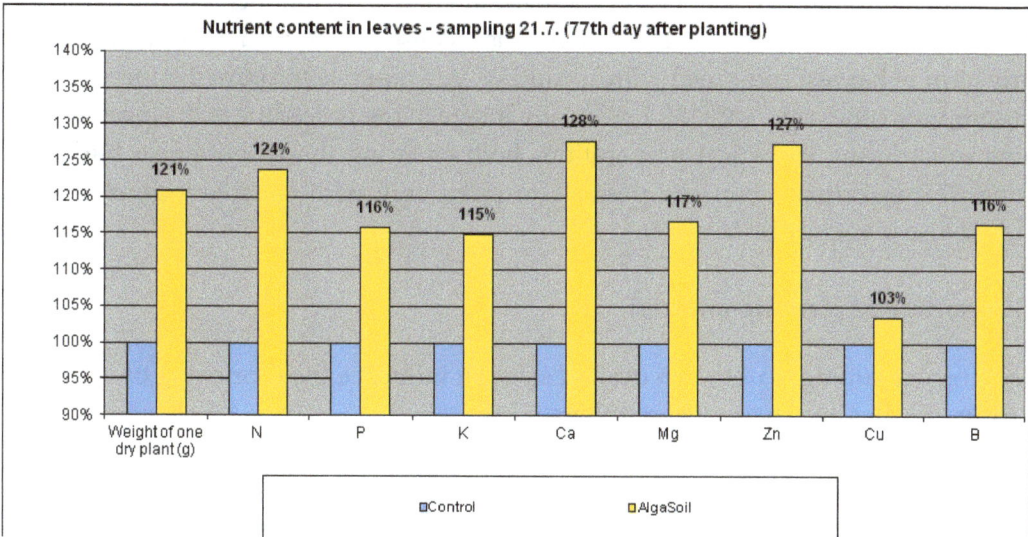

Figure 5. Results of laboratory analyses on the primary nutrients content in the leaves of potatoes on the 77th day after planting [14].

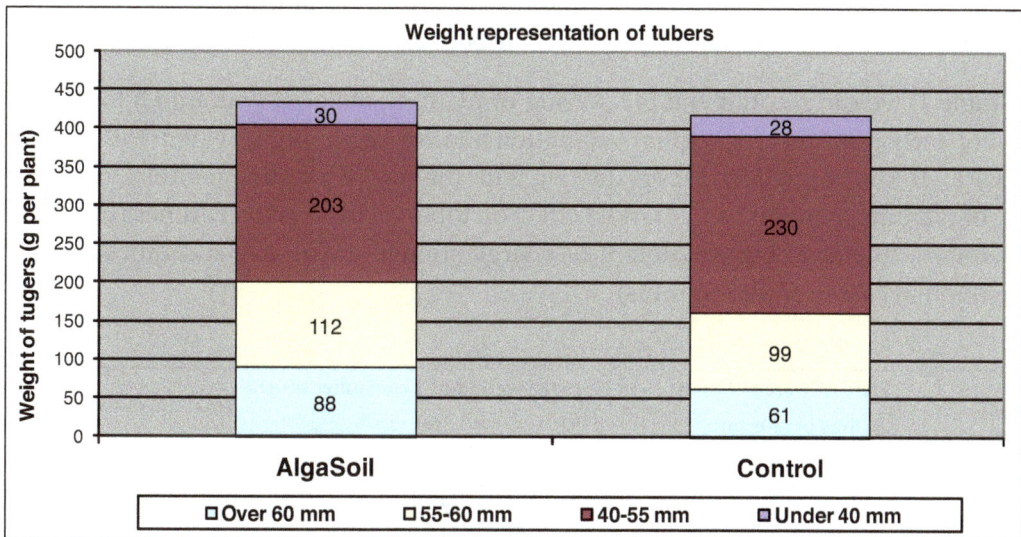

Figure 6. Size analysis under a clump of tubers after treatment with AlgaSoil.

5. Preparation of planting material and planting

The planting material intended for the conditions of organic farming is necessary to sprout or at least to bud. These procedures lead to lower sprout production, which means lower stalks production. This would express as lower tuber setting under the clump, but the tubers would reach the consumption size sooner. So, by these methods, we can increase the earliness and partially anticipate the decrease of production as a consequence of late blight attack. In case of early term of harvest combined with sprouting, it is possible to count the increase of yield of consumption tubers by 7–8% [8]. The disadvantage of the procedure is the increase of work requirements during biological preparations both ensuring the sprouting or budding and planting. The sprouted tubers are possible to plant only with suitable technology (potato planter or disc planting mechanism).

5.1. Size sorting of planting material

Size sorting of planting material on desired size can influence the shortening of vegetation of very early varieties and their yield of tubers.

5.1.1. Experimental verification

In a precise field experiment, three sizes of planting material were compared: variant A (tubers 25 to 35 mm), variant B (tubers 40 to 60 mm), and variant C (65 to 85 mm) with the aim of finding the influence of tuber size on potato yield characteristics. In the experiment, a very early variety called Impala was used. Every variant was set in three repetitions under non-woven fleece textile (Pegas-agro 17 UV) and an uncovered variant was used as control. The harvest and evaluation of yield happened on the 56th to 68th day from planting.

From Table 1, we can recommend big sorting of planting material (variant C) for very early harvest of early potatoes (for regular vegetation and for vegetation covered with nonwoven fabric CFT). It was verified by papers dealing with the size of planting material [5–7] that big sorting of planting material has a marked effect on tuber yield, even on earliness of vegetation (quicker start, thanks to bigger content of energy storage molecules, and quicker ability of regeneration in case of frozen sprouts).

Variant	Yield of ware potatoes (tons per hectare)	Average weight of 1 consumer tubers (g)	Total no. of tubers per plant
		Without cover – control (C)	
AC	12.7[a]	48.9[a]	9.7[a]
BC	15.8[ab]	40.4[b]	13.3[b]
CC	19.3[b]	52.9[a]	13.3[b]
$HSD_{0.05}$	4.02	7.73	2.65
		White fleece textile (FT)	
AFT	17.8[a]	57.4[a]	9.5[a]
BFT	21.2[ab]	60.2[a]	10.2[ab]
CFT	24.5[b]	63.3[a]	12.6[b]
$HSD_{0.05}$	5.91	11.25	2.72

Table 1. Effect of seed tuber size on yield and yield characteristics in the stands that cultivated without cover (C) and cultivated under nonwoven fleece textile (FT) in 2005–2006

5.2. Pre-sprouting

The aim of pre-sprouting is the formation of 15 to 25 mm long, colored, and firm sprouts with basis of roots. It is an intensive procedure, which can hasten emergence, vegetation growth, and even harvest [1]. From the view-point of organic farming, prepared planting materials can ensure quicker emergence of vegetation, which means better concurrence against weed. Quicker emergence also reduces the appearance of black scurf of tubers and stem canker. Pre-sprouting is a suitable procedure to speed up tuber production, and in case of late blight and Colorado potato beetle, tubers are in late state of consumption (pre-sprouting increases the yield assurance).

5.3. Treatment of planting material before planting

In conditions of organic farming, the grower has the possibility to treat potato planting material with allowed agents (this can mainly ameliorate and speed up the emergence of potatoes, than protect it against pests and diseases as in conventional cultivation). For interest, it is possible to specify some preparations, which we have tested (Albit, Amalgerol, Galleko, Special, Polyversum, Softguard). It is possible to apply these preparations on tubers before planting (ultra-low volume pesticide application of tubers in pre-shooting room) or directly during planting on the potato planter. It even partially treated the soil nearby simultaneously [1].

6. Treatment before emergence

The treatment before emergence consists of ploughing and harrowing with full mechanical cultivation. The first operation after planting is blind ploughing after 7 to 10 days. In case of early potatoes, it is suitable for quicker emergence to cover less with soil or to start with harrowing (chain harrow or tine harrow for regulation of emerging weeds in phase of cotydelons). Harrowing also disturbs the soil crust, decreases the height of the top soil above the tubers (meaning warmer through the ridges), so they emerge quicker [8]. With ploughing, we destroy weeds in furrows and on the sides of ridges (it is done most frequently 7 to 10 days after harrowing).

For acceleration of vegetation and early harvest, it is possible to cover the vegetation after planting with white nonwoven fleece textile or perforated foil. The nonwoven fleece textile also provides protection against low temperature, but it limits mechanical cultivation and in case of temperatures higher than 22 °C, plants can be damaged under fleece. In an average of nine years, nonwoven fleece textile probably increased yield of tubers by 23.2% in average of years and varieties in early terms of harvest (ca. 60 days after planting) [9].

It is possible to apply mulching materials on the soil surface (on ridges) to improve soil and nutritional conditions. The main benefits of mulching materials are evaporation regulation, reduction of temperature fluctuation of the soil, and repression of weeds. They can be sources of nutrients and can limit erosion and occurrence of some pests. The right choice of suitable mulching material is important for concrete stand.

The first group are organic (herbal) mulch, such as straw, chopped grass, biomass of intercrops, or other plant material, that can be applied on the ridge surface and usually come directly from the farm. For their application, we recommend manure spreaders, separators of bales of straw, or bedding semi-trailer. The straw is used as mulch mainly abroad. It is easy to store, so it is available during the whole vegetation time [10].

The second big group of mulching materials are plastic products or other waste materials (for example, paper). Considering the origin of plastic and the impact of its application on large-area agriculture, it is necessary to reduce this material and suitably replace it. The use of biodegradable foil or black nonwoven textile can bring certain easement in this area. Targeted processing and recycling of waste paper is possible to produce paper mulching matting with different firmness and durability. The firm VUC Services (www.ekocover.cz) is engaged in this processing and production in the Czech Republic.

In connection with mulch application, it is necessary to mention that mechanical cultivation during vegetation is not possible because of the mulching fabric or foil or it may be limited (in case of plant material). However, past studies imply that the absence of cultivation has no negative effect on tuber yield.

6.1. Application of mulching material

The experiments with herbal mulch, wheat straw, and black textile mulch (weight 50 g/m^2) conducted from 2008 to 2012 brought many answers in the area of temperature change, soil humidity, level of material degradation, biomass of weeds, chlorophyll content in leaves, occurrence of Colorado Potato Beetle (CPB), and Late Blight on tubers and size representation of tuber under clumps [11]. In 2014, we enlarged the experiment by other materials: biodegradable foil and two types of paper matting EkoCover (short-time matting with weight 270 g/m^2 and medium-term matting with weight 800 g/m^2).

6.1.1. Experimental verification

It was found that herbal mulch functions as an isolation and during tracked time decreased the soil temperature by 0.8 °C. Mulch also affected soil humidity conditions when the lowest soaking pressure of soil (that means the highest humidity of soil) was registered at mulching textile. Soil humidity with herbal mulch was in average of years comparable to the unmulched control.

The changed humidity and temperature conditions of soil influence even the nutrient availability in soil [12] and the whole nutritional state of vegetation within it. The source of nutrients for plants can even be its own herbal mulch. The chlorophyll content in leaves was higher by 3.7% in the case of chopped grass applied after planting or before emergence, and it was higher by 2.3% in the case of control (Figure 7). We found the lowest content of chlorophyll in leaves after using black mulching textile and straw (Figure 7 and 8). From known correlation of chlorophyll content and nitrogen content in plants [13, 14], it is possible to deduce that this vegetation had lower nitrogen content in leaves (nitrogen in soil was probably used in straw decomposition not by plants). Other mulching materials (such as paper mulching matting, biodegradable foil) applied after planting (Figure 8) induced lower chlorophyll content in leaves.

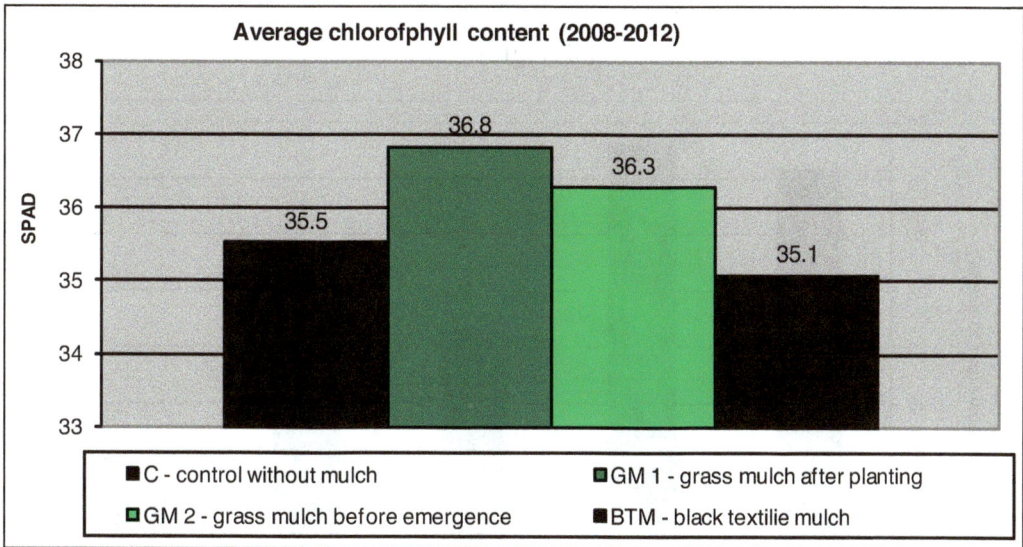

Average chlorofphyll content (2008-2012)

- ■ C - control without mulch
- ◨ GM 1 - grass mulch after planting
- ◨ GM 2 - grass mulch before emergence
- ■ BTM - black textilie mulch

Figure 7. The chlorophyll content (SPAD in units) for each variant of mulch.

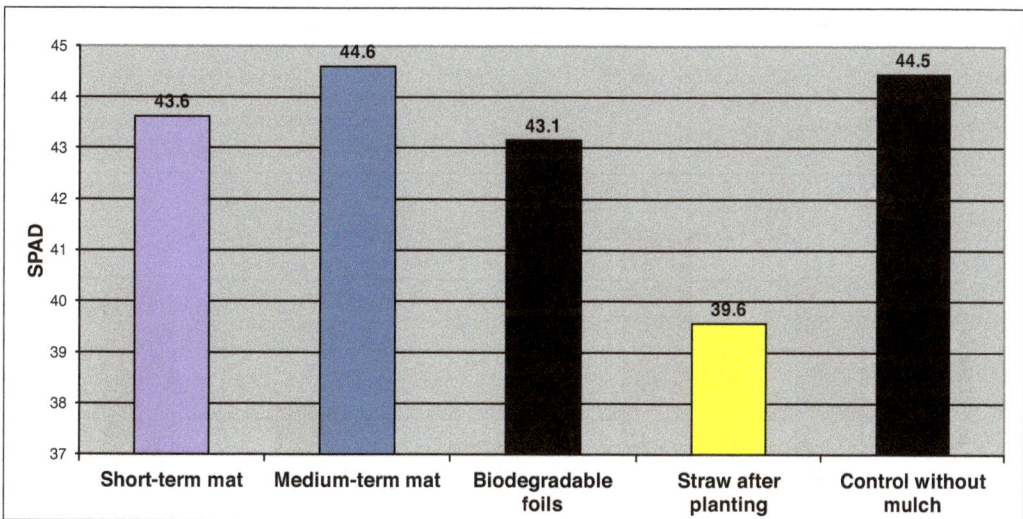

Figure 8. Chlorophyll content in experiments with biodegradable materials (Uhříněves, 2014).

Positive humidity and nutritional conditions affect even growth and biomass of weed and its regulation is ensured only by mulching fabric, biodegradable foil, and paper matting. The application of mulch (or the present weed biomass) is an effective way of soil protection against erosion because the soil is most vulnerable since the planting [15].

The mulch also affects the occurrence of CPB and the following damage of vegetation by the larvae of CPB. Chopped grass reduced the occurrence of CPB (Figure 9) and on the contrary, black mulch textile increased its attack (probably because of higher temperature of soil). The lowest occurrence was found on plots with applied straw. Similarly in 2014, the lowest occurrence of larvae was on straw and foil (Figure 10).

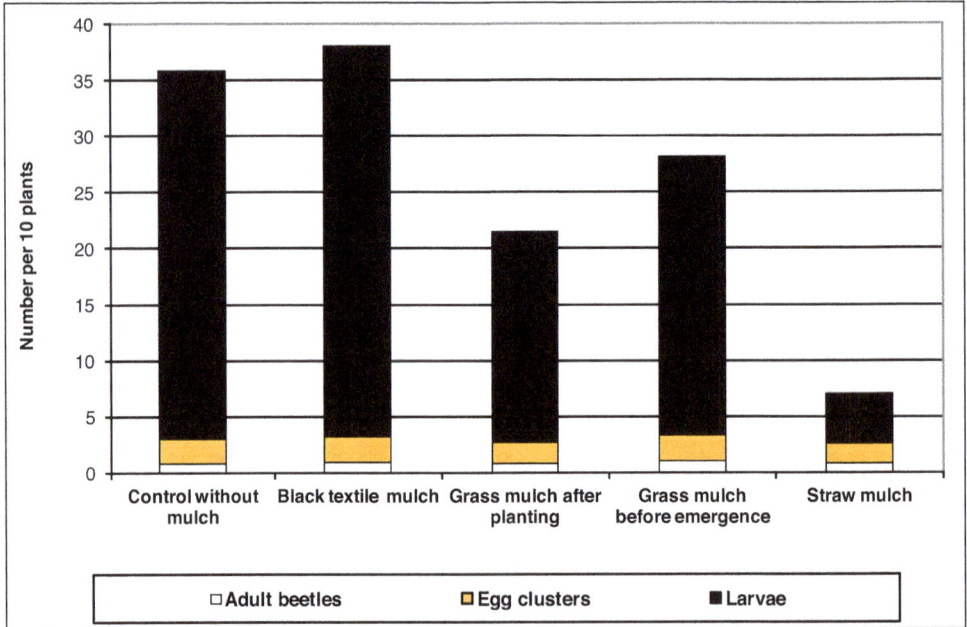

Figure 9. Dependence of the occurrence of beetles, nests with eggs and larvae of CPB on used mulching materials on station Uhříněves (2008–2012).

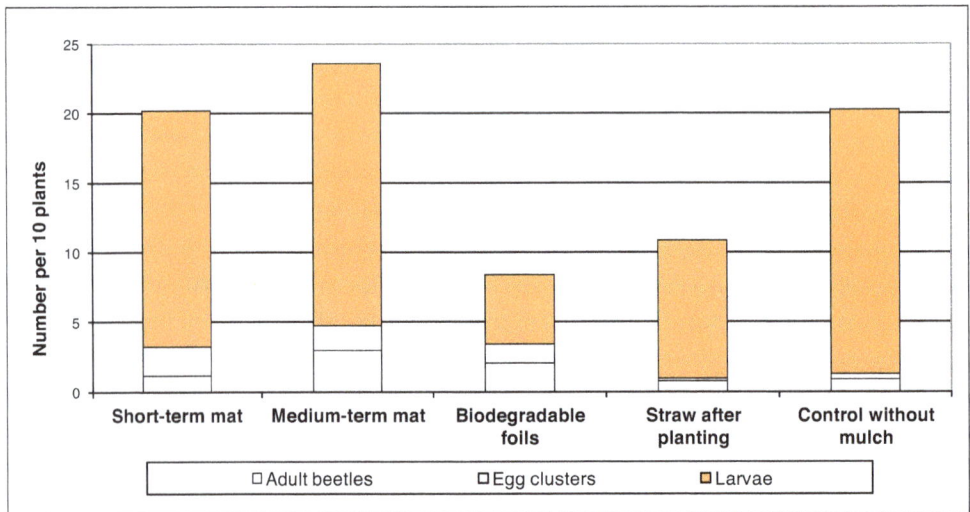

Figure 10. Dependence of occurrence of beetles, nests with eggs and larvae of CPB on used mulching materials (Uhříněves, 2014).

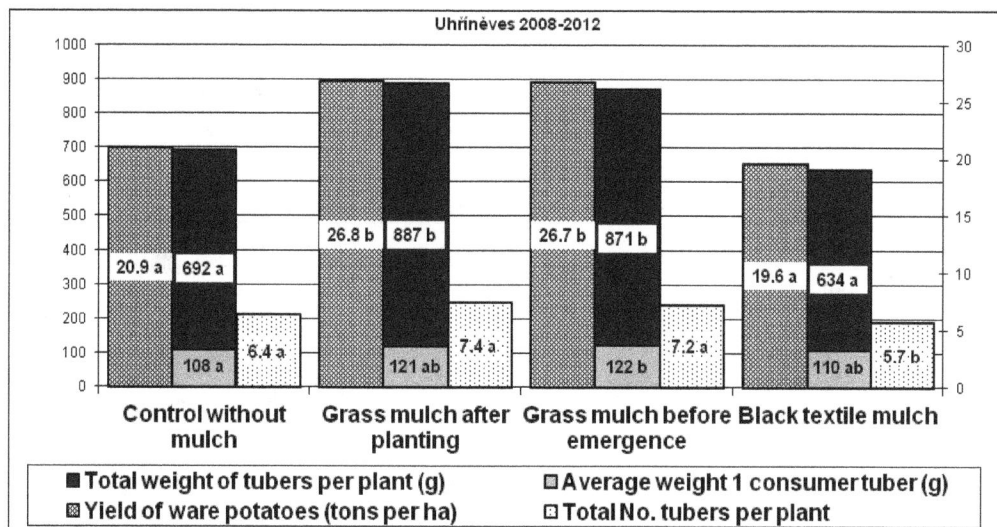

Figure 11. Total weight of tubers, number of tubers, and yield of ware potatoes at various ways to mulching in Uhříněves (different letters for average mean statistically significant differences at the 95% confidence level).

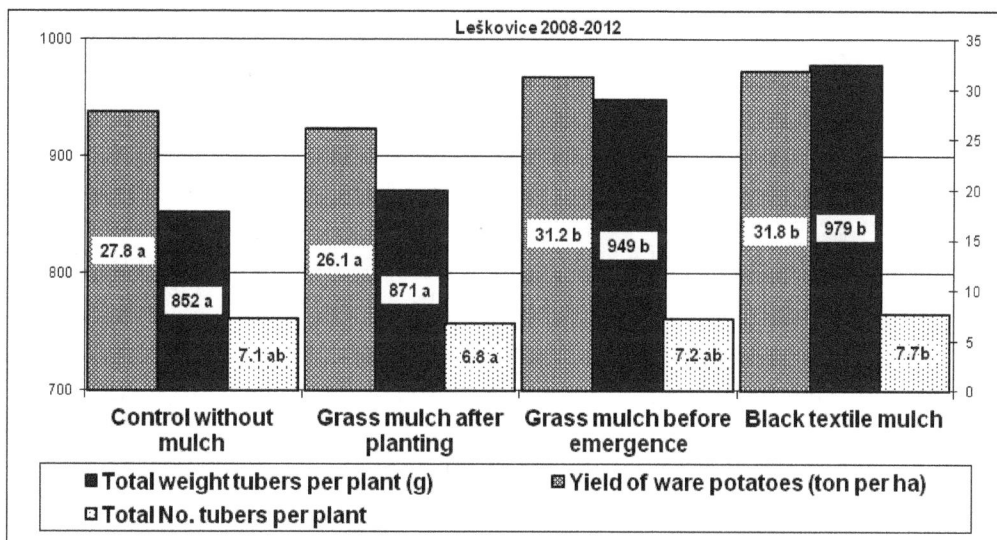

Figure 12. Total weight of tubers, number of tubers, and yield of ware potatoes at various ways to mulching in Leškovice (potato growing region).

The abovementioned factors affect consequent tuber production (Figures 11, 12 and 13). The higher yield of consumption tubers was after the application of chopped grass. Yield of tubers in Uhříněves was lower after the use of black textile mulch than at non-mulched control because of the great attack and damage of vegetation by larvae of CPB. On the contrary, the positive result was achieved with textile mulch on site in the potato processing area where the occurrence of CPB was not high. Black textile mulch positively increased the temperature of the soil and water content in the soil. It produced better conditions for growth and on this site was the highest yield of consumption tubers with textile mulch (higher by 4 t/ha against control).

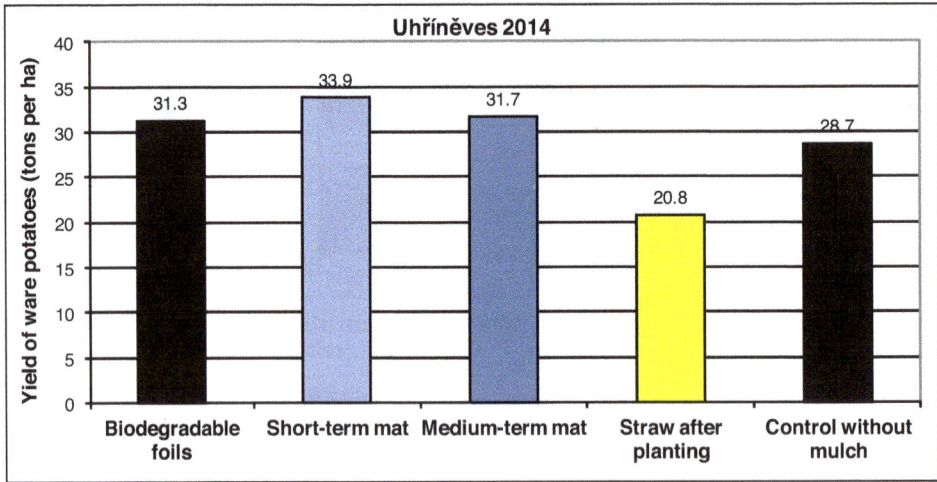

Figure 13. Yield of ware potatoes depending on the selected mulching material.

7. Treatment after emergence

After emergence of vegetation, we continue in mechanical cultivation, which consists of ploughing (eventually the use of weeder) and careful harrowing. Freshly emerged stalk is sensitive on damage, so we should practise harrowing only exceptionally. When the stalk is green and firm, harrowing is possible without great damage in the afternoon hours (when the stalks are withered). In that case, it is beneficial to use tine harrow. It damages stalks lesser than the chain harrow.

According to the need, ploughing (eventually harrowing) is repeated approximately 3 to 4 times until the full canopy closure [8]. The last cultivation intervention should be made until the formation of flower buds when they pile up the ridges as a precaution for transition of late blight from stalk to tubers.

In case the plant height reaches approximately 20 cm, it is suitable to apply (on the leaf or partially also on the soil) supportive preparations (Albit, Alga 600, Alginure, Amalgerol Premium, Ferbiflor, Lignohumate B, PRP-EBV and others).

8. Regulation of pests and diseases

Potatoes can have many diseases (i.e. viral, bacterial, or fungal). For the major part, it is possible to only apply preventive procedures. Direct intervention is possible only in case of fungal diseases.

Potato pests attack mainly stalks and tubers. Some of them are also transferring agents of diseases (for example, aphides transfer plant viruses).

8.1. Late blight (Phythophora infestans)

Late blight is a serious disease on the worldwide scale. If the conditions are favorable, it spreads quickly, and after three weeks, it is able to totally defoliate vegetation [16]. Its regulation in conditions of organic farming is very difficult. The grower must maximally use available preventive methods of pathogen regulation. The assumption is to use known pathogen biology including his weakness.

Varieties of potatoes show marked differences in susceptibility to the late blight. The choice of variety is deciding, because the possibilities of direct crop protection are limited in organic farming. Abroad are already known resistant varieties (Defender, Jacqueline Lee) or varieties with high resistance against it (Sapro Mira, Bionta).

Early term of planting and biological preparation of planting material reduce mainly the risk of yield loss because the later the epidemic shows up (in later stage of plant development), the bigger the tubers are and the lower the losses of yield.

For regulation of late blight, it is possible to use methods that decrease the time of moistening. In case of irrigation need, it is preferable to use the drop irrigation than the spray irrigation (it also saves water). Time-controlled irrigation can markedly decrease the time of moistening. The best time of irrigation is early in the morning, during dew [17]. Unambiguously, it is not suitable to irrigate in late afternoon hours when the stalks cannot dry up before sun-down and usually stay moistall night, which leads to wetting for a very long time and to higher risk of diseases.

The recommended methods of regulation of late blight are suitable organization of vegetation (spacing and row orientation). Orientation of rows is recommended for dominant air circulation. Wide rows (80 to 90 cm) can enhance air circulation and wider rows (90 to 120 cm) prevent canopy closure, which assure longer time of air circulation and makes vegetation dry faster after precipitation. But after, there is lower soil shading and higher concurrence of weeds. Weed occurrence in potato vegetation decreases air circulation and increases the infection risk. In addition, those weeds can be hosts to late blight (*Solanaceae*).

It is also possible to introduce some plants in the vegetation that can reduce the risk of late blight. These new plants form a barrier against the spreading of spores. Some studies mention positive effects of intercropping potatoes with wheat. Potatoes are planted diagonally to dominate air circulation and wheat is sowed in the furrows. Another alternative method verified in the project Blight-MOP with positive result was alternate (band) cultivation of varieties resistible and sensitive to the late blight on one site or cultivation of more varieties in one row. This mixture of varieties can improve control over pathogen, but induce practical problems with harvest and variety separation [18].

Balanced plant nutrition including microelements decreases the possibility of late blight infection of potatoes [1]. Overdose of nitrogen fertilizer forms less tubers and lots of stalks that dry up slower, which increases the infection risk. More resistant are mature "older" stalks [16] well-supplied with potassium [1].

In case of occurrence of late blight in the vegetation (when preventive methods did not work), it is possible to alternatively approach the destruction of the first infected plants on site. It can stop or slow down the spreading of the disease to the rest of the site. We have to eliminate not only the visibly ill plants, but also the plants around the focus point because they may be infected though without any symptoms. The appearance of symptoms takes around three days to one week (depending on environmental conditions). Results of these methods are the elimination of many apparently healthy plants, which are enclosed by the infected plants. For these purposes, it is possible to use, for example, a propane-butane burner, which can ensure the destruction of spores.

Opinions on the use of preparations on the basis that copper is markedly different (grower to grower, state to state) is mainly dependent on legislation. In some states, copper fungicide was limited. According to the EU, they determined a boundary of 6 kg of Cu/ha/year. In Scandinavia, copper fungicides cannot be used at all. Growers there are trying to use alternative products, but with smaller success. In present conditions, the ban of copper fungicide could destabilize the production of organic potatoes because there are no other effective alternatives for blight regulation.

In our experiments from 2009 to 2011, solutions of plant and animal origin were tested and supplemented with five hopeful commercial preparations (Figure 14). First, preventive spraying was always done before occurrence of blight, and consequent treatment was done according to prognostics and signalization. The site, where the experiment occurred, was typical for lower blight attack on stalks and tubers, so even the use of alternative spraying had satisfactory results compared with copper fungicide. We also observed mild phytotoxicity of preparation with the extract from walnut tree (*Juglans regia* L.), which probably had an effect on tuber yield.

Treatment	Late Blight on the leaves (1–9)	Tuber blight		Tuber yield (tons per ha)
		Number (%)	Weight (%)	
Kuprikol 50	7.6	0.4	0.3	26.1
5% solution of biomilk	7.5	0.6	0.5	23.3
10% extract from *Juglans R.*	7.6	1.0	0.9	22.6
MycoSin VIN*	7.6	1.1	0.3	28.0

Note:* Only years 2010–2011, 9 points – without challenge

Table 2. Incidence of Late Blight on the leaves and tubers of potato (expressed in % of infected leaves and infected tubers)

Another comparison of commercial preparations is represented in Figure 14. Surprisingly, the best results on blight regulation were observed with preparations against Colorado potato beetle (Neem Azal T/S and safety net). It affirmed the recent finding that regulation of CPB in organic farming (regulation of leaf damage) has a positive effect on the decrease of blight in potato vegetation.

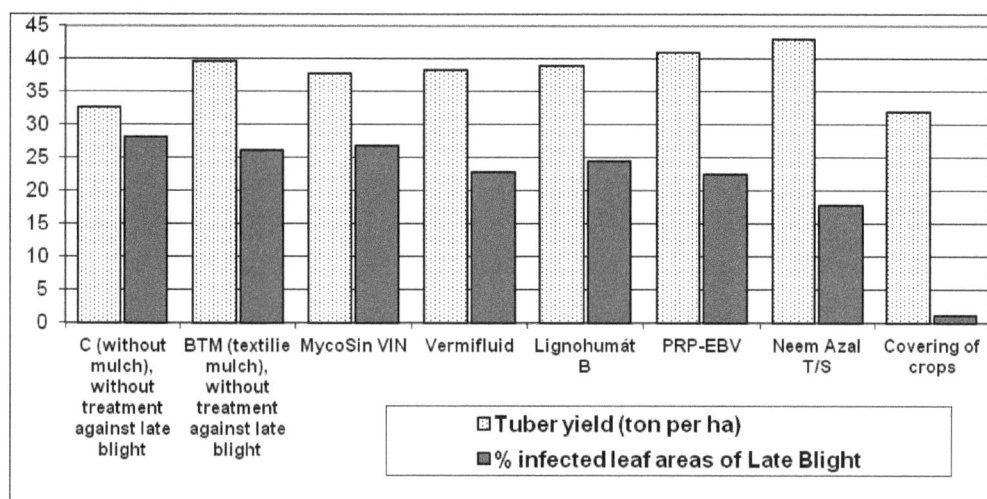

Figure 14. Results of applications support preparations in average varieties (Monika, Jelly, and Red Anna) on station Uhříněves (2009–2011).

8.2. Colorado potato beetle (Leptinotarsa decemlineata)

CPB is a pest of potatoes, which after overpopulation induce serious damage of vegetation and decrease of tuber yield [1]. The biggest damage is caused by its larvae. Their overpopulation can lead to clean-eating, leading to the destruction of vegetation. This pest should not be undermined.

From preventive precaution, it is possible to recommend pre-sprouting and early planting, not place the potatoes on near-by sites (easily admissible for beetles), aim for support of natural enemies (lady-bugs, heteroptera, earwig, and birds such as blackbird, pheasant, or partridge), and application of mulch. From variety experiments are some possible different attacks (attractiveness of varieties for Colorado beetle). The deciding factor can be the content of glycoalcaloids or trichomes on leaves.

Direct crop protection on large area consists of applications of biological insecticide. Currently registered in the Czech Republic are two effective substances: azadirachtin (in Neem Azal T/S) and spinosad (in Spintor). In some states, it is possible to use biological preparation Novodor FC on the basis of bacteria *Bacillus thuringiensis* var. *tenebrionis*.

On smaller areas, it is possible to use labor-intensive way of hand collecting (mainly of spring beetles), which aims to prevent the laying of eggs. Uniquely, it is possible to find special shakers or blowers (eventually vacuums), but they are usually homemade machines or prototypes.

9. Preparation of harvest and harvest methods

Removing stalks happens usually early in organic farming because of late blight (with the removal of stalks, we follow the regulation of inoculum and spread of infection on tubers). In case of very early potatoes, we remove stalks mainly for simplification of harvest and hardening of the peel (in this case 2 to 3 weeks before planed harvest). To remove the stalks, we use a mechanical stalk crusher in organic farming.

In case of production of planting material, stalk removal is necessary and unavoidable mainly from the point of view of viral regulation (eventually the pass of aphids). A more efficient procedure (mainly with planting material) is thermic removal of stalks (fire, vapor, or nitrogen).

On certain conditions it is possible to use even the tweezers of stalks (only with erect vegetation on consistent soil so it would not tear out the tubers). These machines are not available in the Czech Republic, so there are not even used [19].

10. Conclusion

Existing knowledge and experiences in the technology of cultivation of organic potatoes are continuously innovated and specific issues of growers are addressed. Especially valuable are the findings in the field of soil treatment and processing as they affect the soil state and the soil edaphon, which has an irreplaceable role in the system of ecological agriculture. Adequate soil treatment and application of organic materials, combined with biological preparations, have positive impact on the nutritional state of vegetation and are effective ways of how to balance nutrients in organic farming. In our experiments, the nutritional state of vegetation and the tuber yield improved by using soil preparations of Azoter (yield increased by 1.1 t/ha) and AlgaSoil (increase by 0.9 t/ha). The growers solved the nutrient deficit found during vegetation only marginally. Even here, the supply is growing and the organic grower already can apply liquid or organomineral fertilizers with quick nitrogen effect on the basis of actual nutritive state of plants. Another big group of preparation is the so-called supplemental plant preparations. We had the chance to verify some preparations of this group (Albit, Alga 600, Amalgerol, Lignohumate B, PRP-EBV, or Softguard) with positive results. Another benefit of these preparations is their possible effect on the health of plants.

The health, even the tuber yield, of potatoes is possible to influence with some other operation such as the choice of variety, size sorting of planting material, and treatment of plant material. The use of greater size sorting and of tubers of overplanting size increased the tuber yield in combination with early harvest. It is necessary in organic farming to perform biological preparation of planting material (pre-sprouting) because of late blight. Consequential growth of roots and vegetation vitality is possible to support with treatment of planting material before planting or application during planting (on potato planter).

The protection of soil and soil life is also important. The application of mulching material on top of ridges can help in this area (also as anti-erosion precaution). Another benefit of mulch is that it can be used in the regulation of CPB or aphids (mainly if we use herbal mulch or buffer strip), weed regulation (weed biomass was regulated by black mulching fabric and partially by grass mulch applied before emergence), the possibility of temperature and humidity regulation, and also the increase of tuber yield. The tuber yield was largely affected by concrete use of mulching material (the right choice of mulching material unrolls from concrete site and soil conditions).

Initial treatment of potato vegetation happens according to concrete environmental conditions of the year and the grower's experiences. We gain many valuable results about plant extracts of *Azadirachta indica* L., eventually Neem Azal and other plant extracts (*Juglans regia* L., *Pelargonium zonale* L.) as protection against late blight and CPB. However, they are not usable in practise because of their changing effectivity. The main regulation procedure includes: 1. choice and use of resistant varieties; 2. pre-sprouting of planting material and early planting; 3. suitable irrigation regime; 4. interchange of crops; 5. removal of stalks or use of copper fungicide.

Aimed liquidation of stalks stops not only the blight spread, but also its transition to the soil and on tubers. Another area using stalk removal is the regulation of maturation (tuber size) and regulation of viruses (in propagation vegetation). Experimental results indicate, that even after stronger pass of aphids, it is possible to use preventive methods (early varieties, pre-sprouting, early planting, buffer strip, or mulching) in organic farming and regulate the occurrence of viral diseases. Production of planting material is possible even in conditions of organic farming. It demands good knowledge and maximal usage of all regulation methods and procedures.

Acknowledgements

This publication are from results gained from solving the projects MSM6046070901, MSN 590 G4, CIGA 20062005, MZe ČR č. QH82149.

Author details

Petr Dvořák*, Jaroslav Tomášek, Karel Hamouz and Michaela Jedličková

*Address all correspondence to: dvorakp@af.czu.cz

Czech University of Life Sciences Prague, Faculty of Agrobiology, Food and Natural Resources, Kamýcká, Prague – Suchdol, Czech Republic

References

[1] Vokál B. 2004. Technologie pěstování brambor: (rozhodovací systémy pro optimalizaci pěstitelských technologií u jednotlivých užitkových směrů brambor). Praha: Ústav zemědělských a potravinářských informací, 91s. ISBN 80-727-1155-5.

[2] Dvořák P., Bicnová E. 2007. Brambory v systému ekologického zemědělství. Sborník Ekologické zemědělství 2007. ČZU Praha, 6-7.2.2007: 131-133.

[3] Diviš J., Valeta V. 2006. Která odrůda bramboru je vhodná. Zemědělec, 7: 42.

[4] Vos J., Born M. 1993. Hand-held chlorophyll meter: A promising tool to assess the nitrogen status of potato foliage. Potato Res., 36: 301-308.

[5] Diviš J., Bárta J. 2001. Influence of seed-tuber size on yield and parameters in potatoes. Rostlinná Výroba, 47(6): 271-275.

[6] Votoupal B. 1964. Velikost sadbových hlíz. Rostl. Výr., 10: 1033-1042.

[7] Wiersema S.G. 1989. Comparative performance of three small seed tubers and standard size seed tubers planted at similar densities. Potato Res., 32: 81-89.

[8] Dostálek P., Hradil R., Křišťan F., Škeřík J. 2000. Bulletin ekologického zemědělství č. 18, téma – Brambory. PRO-BIO Šumperk: 24s.

[9] Dvořák P. 2008. Vliv agrotechnických faktorů na ranost sklizně a spotřebitelskou jakost brambor. Dissertation thesis. ČZU Praha: 140s.

[10] Dvořák P., Tomášek J., Hamouz K., Cimr J. 2013. Ověřený postup v ochraně půdy a porostů brambor. Agricultura - Scientia - Prosperitas. Intenzifikace rostlinné výroby a trendy pěstitelských technologií. Praha: 55-60. ISBN: 978-80-213-2351-3.

[11] Dvořák P., Tomášek J., Hamouz K., Mičák L. 2013. Začlenění systému povrchového mulčování do technologie pěstování brambor. Certifikovaná metodika. ČZU Praha: 32s.

[12] Fang S.Z., Xie B.D., Liu D., Liu J.J. 2011. Effects of mulching materials on nitrogen mineralization, nitrogen availability and poplar growth on degraded agricultural soil. New Forests, 41: 147-162.

[13] Gianquinto G., Goffart J.E., Olivier M., Guarda G., Colauzzp M., Costa L.D., Vedove G.D., Vos J., Mackerron D.K.L. 2004. The use of hand-held chlorophyll meters as a tool to assess the nitrogen status and to guide nitrogen fertilization of potato crop. Potato Res., 47(5): 35-80

[14] Hašková P. 2014. Jaký vliv má organické granulované hnojivo? Agromanuál, 8: 29.

[15] Truman C.C., Shaw J.N., Reeves D.W. 2005. Tillage effects on rainfall partitioning and sediment yield from an ultisol in central Alabama. J. Soil Water Conserv., 60(2): 89-98.

[16] Stone A. 2012. Organic Management of Late Blight of Potato and Tomato (Phytophthora infestans) [Online]. eXtension. http://www.extension.org/pages/18361/organic-management-of-late-blight-of-potato-and-tomato-phytophthora-infestans (downloaded 18.9 2014).

[17] Kirk W., Wharton P., Hammerschmidt R., Abu-El Samen F., Douches D. 2007. Late Blight [Online]. Michigan State University Extension Bulletin E-2945. East Lansing, MI. http://www.potatodiseases.org/lateblight.html (downloaded 18.9 2014).

[18] Leifert C., Wilcockson S.J. 2005. Blight-MOP: Development of a systems approach for the management of late blight (caused by Phytophthora infestans) in EU organic potato production. University of Newcastle, UK.

[19] Bioinstitut, 2007. Praktická příručka č. 4 Biobrambory - Jak ekologicky vypěstovat kvalitní brambory. Bioinstitut, Olomouc: 23s.

8

The Role of Biological Diversity in Agroecosystems and Organic Farming

Beata Feledyn-Szewczyk, Jan Kuś, Jarosław Stalenga, Adam K. Berbeć and Paweł Radzikowski

Additional information is available at the end of the chapter

Abstract

Ecosystems are the basis of life and all human activities. Conservation of biological diversity is very important for the proper functioning of the ecosystem and for delivering ecosystem services. Maintaining high biodiversity in agroecosystems makes agricultural production more sustainable and economically viable. Agricultural biodiversity ensures, for example, pollination of crops, biological crop protection, maintenance of proper structure and fertility of soils, protection of soils against erosion, nutrient cycling, and control of water flow and distribution. The effects of the loss of biodiversity may not be immediately apparent, but they may increase the sensitivity of the ecosystems to various abiotic and biotic stresses. The combination of biodiversity conservation with profitable food production is one of the tasks of modern sustainable agriculture that faces the necessity of reconciling the productive, environmental, and social goals. As further intensification of production and increase in the use of chemical pesticides, fertilizers, and water to increase yields are increasingly criticized, global agriculture is looking for other biological and agrotechnical methods in order to meet the requirements of global food production.

Keywords: Biological diversity, ecosystem, agroecosystem, ecosystem services, organic agriculture

1. Introduction

In compliance with the Convention on Biological Diversity (CBD), adopted in Rio de Janeiro in 1992, biological diversity is the variability among living organisms inhabiting all environments and ecological systems [1]. Biodiversity may therefore be considered at genetic, species, and ecosystem levels. According to Clergue [2], biodiversity is a very complex issue. In agroecosystems, it serves three basic functions: genetic, agricultural, and ecological functions. The first function of biodiversity involves maintaining species gene pool, in particular, the endangered ones. The second function, connected with agricultural activity, contains increasing the resistance of agroecosystems to abiotic and biotic stresses, as well as maintaining their productive role. Biodiversity has also ecological functions, for example, creating habitats with different flora and fauna species that have specific significance in agroecosystems.

The loss of biological diversity is one of the most important problems of the world and a threat to our civilization. The destruction of primary ecosystems, intensive farming, urbanization, and also infrastructure development cause depletion and weakening of the stability of ecosystems. Agroecosystems are the most at risk of losing biological diversity [3].

During the last decades, worldwide losses of biodiversity have occurred at an unprecedented scale and agricultural intensification has been a major driver of this global change [4]. The dramatic land use changes include the conversion of complex natural ecosystems to simplified ecosystems and the intensification of resource use, including application of more agrochemicals. The evaluation of ecosystems in the UK has shown a significant loss of biodiversity during the recent 50 years. Sixty-seven percent of 333 plant and animal species on agricultural lands have been endangered, mainly due to the intensification of farming [5].

The industrialization of agriculture has caused, directly and indirectly, a dramatic impoverishment of the fauna and flora compared to the situation a century ago [6–9]. This has contributed not only to the current biodiversity crisis in Europe as whole, but also to the decline in ecosystem services such as crop pollination and biological pest control [8]. As a result, the protection of farmland biodiversity has become a key issue in the EU and national agricultural and environmental policies, and large amounts of research and funding are devoted to biodiversity conservation, such as agri-environment schemes [3, 10–11].

Despite the commitment made by the Parties to the Convention on Biological Diversity to reduce the rate of biodiversity loss by 2010, global biodiversity indicators show continued decline at steady or accelerating rates, while the pressures behind the decline are steady or intensifying [12]. The main objective of the EU Biodiversity Strategy to 2020, which was adopted in 2011, is to maintain and strengthen ecosystems and their functions, and foster sustainable development of agriculture and forestry [13]. Biological diversity should also be preserved due to economic factors. Maintaining a high level of biological diversity makes agricultural production and the related activities more sustainable, which in turn, significantly affects human activities [14–15].

Biodiversity in agriculture can be perceived on two levels: the first is related to the diversity of species and cultivars, the breeds of farm animals, so the obtained "products"; and the second is related to the biodiversity connected with agricultural production, such as the diversity of plants and wild animals that accompany the crops, as well as the diversification of the agricultural landscape.

2. The role of traditional species, cultivars, and traditional animal breeds in maintaining biological diversity

The progress in agriculture has led to the situation that in the recent 100 years, approximately 75% of genetic resources have been lost due to the transition of farmers from growing traditional, local cultivars of lower productivity and replacing them with intensive cultivars. Although in the world there are at least 12 thousands of edible plant species, humans use only 150 to 200 of them, and 75% of food products around the world are produced from only 12 species of plants and animal species. The three main species of plants such as rice, maize, and wheat provide about 60% of the energy consumed by humanity. Such a low diversity is a major issue to food safety. From the point of view of the conservation of biodiversity and human health, we should promote traditional and local species and cultivars of plants, as well as old breeds of animals [16].

The most appropriate way of protecting genetic resources of plants is their conservation in situ in the regions strictly related to their origin. This type of protection allows us not only to preserve a given form in its place of origin, but also to continue its cultivation and selection in the traditional way. The protection of genetic resources of crops, in addition to the primary task of maintaining biodiversity, has also practical aims of delivering rich genetic material for further breeding [6].

Old and local cultivars of crops are distinguished by unusual qualitative characteristics (e.g., good taste, favorable chemical composition), low technological requirements, better adaptation ability to environmental conditions, resistance to pests and diseases, and reliable yields. The cultivation of old cultivars and forms is often connected with using environmentally friendly production systems, such as organic farming. Old varieties are usually cultivated on a limited area, at a local or regional level. In Poland, we cultivate the tradition of growing old and local cultivars of tomato, cucumber, onion, carrots, beans, pumpkin, vetch, and many other orchard fruits and vegetables. In recent years, the rapidly-developing low-input methods of farming promotes a wider use of old and local cultivars of plants, as well as old plant species, such as spelt wheat, emmer, einkor wheat, and their processing on the farm [6].

Traditional orchards, also called backyard orchards, are of great importance for plant genetic resources. They usually satisfy only the needs of their owners and their family, unlike the commercial orchards where the production of which is destined primarily for sale. Traditional orchards became a characteristic element of the landscape of the Polish countryside. Due to the longevity of the trees, they have survived to this day. They are supported by an agri-environment scheme in Poland [17].

Native animal breeds are very important due to the role they played in the history of the development of the regions from which they originate. Due to their ecological, landscape, ethnographic, and socio-cultural functions, they must be regarded as evidence of tradition and culture of local communities, and preserved for future generations. The conservation of genetic variability guarantees a secure future of livestock production and helps maintain a healthy livestock [6].

3. The role of wild flora and fauna diversity in agroecosystem

In intensive conventional farming, special attention is paid to the negative aspects of wild flora in agrocenoses (called weeds), as they cause yield losses. Since the 1990s, however, due to the promotion of the concept of sustainable agriculture, the importance of wild plants growing on fields has been underlined. They have started to be perceived not only as competitors to arable crops, but also as an element that increases the biodiversity in agroecosystems [18–20].

Currently, the tendency in weed control is to limit the number of weeds to such a level that do not cause significant yield decreases. Such an approach is consistent with the objectives of sustainable agriculture, and particularly promoted in the system of organic farming. The harmfulness of weeds is not the same in all agrocenoses and depends on: the species and its biology, their abundance, competitive ability, the type of agricultural culture and the purpose of cultivation, as well as the soil type, weather, and agrotechnical factors [21].

The results of the research indicate a positive influence of wild flora in preserving overall biodiversity of agroecosystems [20, 22]. Elimination of wild plants from plant canopy, and thus weakening their reproductive potential interferes with the processes occurring in soil and relations between flora, fauna, and microorganisms [23]. Studies have shown that the decrease in the number of weeds as a result of the intensification of agriculture in Finland, Germany, Denmark, and the UK caused a decline of the populations of birds, pollinators, and other insects on agricultural areas [20, 22, 24]. The results of the monitoring of common breeding birds, which have been conducted in the UK since the 1990s and in Poland since 2000 indicate that the decrease in the number of the species such as tawny pipit, goldfinch, hoopoe, and lapwing, following the intensification of agriculture and the reduction in the diversity of weed flora [25]. The seeds of weeds, especially from the *Polygonaceae, Chenopodiaceae*, and *Poaceae* families, such as *Chenopodium album, Polygonum aviculare, Echinochloa crus-galli, Rumex obtusifolius*, and *Stellaria media*, are important food components for many bird species [20, 26].

Weeds constitute the source of food, as well as the habitats for animals, including useful, pollinating insects [15]. The nectar and pollen producing plants include: *Anthemis arvensis, Cirsium arvense, Centaurea cyanus, Chenopodium album, Consolida regalis, Taraxacum officinale, Papaver rhoeas*, and *Sonchus arvensis* [20–21]. Many common weed species are significant for the maintenance of the population of valuable beneficial invertebrates (pest predators and parasites), thus supporting the natural pest control [20].

Providing pest control is one of most important functions of biodiversity. There is a significant importance of predatory arthropods in agroecosystems. Many species of invertebrates are specialized in eating aphids and other pests. Others are generalist predators such as spiders or ground beetles. One of the most important natural enemies of pests are spiders. Almost all known species of spiders are predators. Many species are common in crops. The most effective in pest control are species families *Licosidae, Linephidae, Salticidae, Tetragnatidae, Clubionidae*, and *Araneidae* [27]. An important feature of spider biology is its resistance to long periods of hunger when a prey is absent. On the other hand, when prey is in abundance, they can consume a huge amount of it, often killing more prey than they can actually eat [28]. Another very important taxa is *Coleoptera*. There are many species of *Coleoptera*, that are generalist predators feeding on aphids and other pests. In an agroecosystem, the beetle families *Carabidae*, *Staphylinidae*, *Coccinellidae*, and *Cantharidae* are the most important invertebrates. The best known natural enemies of aphids are ladybirds *Coccinellidae* and ground beetles *Carabidae* [29]. Predatory beetles are more common in organic crops and in diverse landscapes [30]. They are also not dependent on pest population density, while specialist natural enemies are. They are also present on the field before pest population has developed. There are more generalist predators that can control the population of pests. These are insects such as bugs *Hemiptera*, robber flies *Asilidae*, wasps, and ants *Hemiptera*. More specialized in aphid control are parasitic wasps *Apocrita-Parasitica*, hoverflies *Syrphidae*, lacewings *Chrysopidae*, and *Hemerobiidae*. Both types of natural enemies are effective in controlling aphids, but they affect them in different ways. Generalist predators limit pest population, but doesn't eliminate all individuals so there is still a possibility to rebuild pest population. Specialists influence pest population slowly, preventing the increase in the population [31]. Diversity and activeness of natural enemies depends on the type of crop, diversity of landscape, and system of farming.

High plant species diversity increases the diversity of soil microflora and microfauna, including the organisms that are antagonistic against crop pathogens [32]. Certain wild flora species repel the crop pests or they act as trap plants for pests (e.g., *Chenopodium album* for black bean aphids). The allelopathic potential of many weed species has a stimulating or inhibiting effect on the development of crops and the presence of other weeds [21]. A large variety of flora and fauna is increasingly perceived as a valuable part of the agricultural landscape, especially in countries where intensification of agricultural production has led to a significant reduction of biodiversity of agroecosystems [19].

4. Biodiversity in the ecosystem services concept

Ecosystem services have become a top research issue in ecology, natural resource management, and policy [33]. Ecosystem services can be defined as the benefits that humans obtain from ecosystems [34].

In the report of Millennium Ecosystem Assessment [35], ecosystem services were divided into four basic types:

- provisioning (production of food, production of other raw materials such as wood, fuel, water supply, and others);

- regulating (regulation of air composition, climate, extreme phenomena, contamination, and biological processes);

- supporting (circulation of elements, primary production, soil formation, habitat function, hydrological cycle);

- cultural (recreational, aesthetic, cultural, and educational functions).

Biodiversity plays a major role in each group of these ecosystem services. It is crucial for the functionality, stability, and productivity of every ecosystem. In dynamic, agricultural landscapes, only a diversity of insurance species may guarantee resilience (the capacity to reorganize after disturbance) [8]. The species that occur in agrocenoses differ in terms of their potential value and input into the ecosystem services [15, 36]. Thus, increasing the diversity of species richness increases the probability of the total pool containing a species that will significantly affect the functioning of the ecosystem.

Biodiversity and ecosystem services are complex issues, which is reflected in many different interpretations of the significance of biodiversity to the ecosystem. The connections between biodiversity and ecosystem services are perceived differently by different authors [37]. Some authors even treat these concepts as one, which means that if the ecosystem services are managed properly, biodiversity will be preserved and vice versa ("ecosystem services perspective"). However, others claim that biodiversity is one of the ecosystem services and the conservation of the diversity of wild species, especially the endangered ones, is one of the goods that the ecosystem should deliver ("conservation perspective").

According to Fischer and Young [38], in biodiversity, everything is connected and contained in the same environment, but with no hierarchy. Mace et al. [37] suggest that the role of biodiversity in ecosystem services should be put into some order by assuming that different relationships exist at different levels of the hierarchy of ecosystem services. Following this concept, biodiversity may be the primary regulator of the ecosystem processes, as well as the final product and ecosystem service and good itself.

Biodiversity is considered one of the provision services that can supply: genetic resources for breeding new, more useful cultivars of plants or animal breeds; new active substances for medicine and pharmacology; or new ornamental plants [37]. Biodiversity in ecosystems determines most of the basic functions of the ecosystem, such as the distribution and circulation of elements in soil or the resistance of the ecosystem to pests and environmental conditions. It is generally considered that a more diverse ecosystem is a more stable ecosystem. The results of the studies indicate that an increased biodiversity at a given trophic level positively affects the productivity of this trophic level [39].

Ecosystems with high biological diversity provide many ecosystem services that concern, among others, provision of food, maintenance of pollinators, and biological control of pests [8, 15]. Pollination is one of the ecosystem services that are of special importance for humans. Recent studies estimate that 87 of major arable crops and 35% of the world crops are pollinated by animals [40]. The diversity of pollinators is essential for maintaining the provision of the services that Costanza et al. [34] evaluated at $14/ha/year. According to other authors, it amounts to $100 billion a year around the world [41]. The loss of biodiversity of agroecosystems, caused by the intensification of agricultural production and the loss of habitats, negatively affects the service of pollinators, which causes yield decrease [42].

The studies on the influence of biodiversity on ecosystem functions are difficult due to the complexity of the relationships within the ecosystem, the impact of agricultural production systems, and landscape. It is also difficult to generalize the results obtained in the given ecosystem over other ecosystems [43].

Meta-analysis carried out by Balvanera et al. [39] indicates that most of the published works show a positive influence of biodiversity on the functioning of ecosystem, the strongest at the level of communities. Costanza et al. [44] found a positive impact of biodiversity on the productivity of ecosystems in North America. According to these authors, 1% of the changes in biodiversity affects 0.5% of the changes in the value of ecosystem services. The research carried out in Europe provided evidence for the positive impact of biodiversity on the productivity of grasslands [45]. Lavelle et al. [46] pointed to the positive impact of diversity of soil organisms on plant productivity in agricultural ecosystems. Hillebrandt and Matthiessen [47] believe that the functioning of the ecosystem is dependent not only on biodiversity, measured by the number of species, but most of all, on species composition, and the abundance of individual species and functional groups. A recent review of the scientific literature concluded that most reported relationships between biodiversity attributes (such as species richness, diversity, and abundance) and ecosystem services were positive [48]. Despite rich evidence on the existence of the connection between biodiversity and ecosystem functioning, some authors still question this relationship [8, 49–50].

The protection of certain target species is the most socially recognized role of biodiversity, while its indirect role in processes occurring in ecosystems (such as the cycle of elements) is little known by a wider audience [37]. A higher perspective needs to deliver additional arguments for the protection of biodiversity, apart from the traditional arguments, connected with the protection of rare and charismatic species.

Authors of the report from ecosystem evaluation in the UK found that at present, we are not able to fully assess the relationship between biodiversity and ecosystem services that it provides [5]. Changes in the extent and condition of habitats may significantly affect biodiversity ecosystem services. Intensification of agriculture has caused agricultural production, along with provision services, to significantly increase, but at the same time, there was a reduction in the diversity of the landscape, the increase of soil erosion, the reduction of soil quality, and the decrease in the populations of birds and pollinators. Changes in ecosystems may have a positive or negative impact on human welfare. For example, the conversion of natural ecosystems into agricultural production areas increases farmers' income, but at the same time, decreases habitats for recreation and the threat of atmospheric phenomena. According to the authors of the report [5], these types of assessments, in addition to economic values, should also take into account human health and social values.

Until now, ecosystem services were regarded as public goods, not as a market product that has a monetary value. According to some authors, the lack of valuation is the main cause of the degradation of ecosystems and loss of biodiversity [3]. If we want to maintain our environmental safety, we have to "measure" ecosystems and biodiversity. The article of Costanza et al. [34], "The value of the world's ecosystem services and natural capital", published in Nature in 1997, was a breakthrough study in the subject of ecosystem services valuation. The authors assessed the value of 17 basic services produced by ecosystems all over the world. They evaluated them at $33 billion per year, so almost twice the amount of the gross national product of the USA ($18 billion). The concepts of ecosystem services flow and natural capital stocks are increasingly useful ways to highlight, measure, and value the degree of interdependence between humans and the rest of nature [51]. Economic assessment of the value of the services provided by the environment is difficult, time-consuming, and flawed. The valuation of each group of ecosystem services should be performed using different methods [52–53].

5. The impact of different agricultural systems on biodiversity

One of the most important factors affecting the agroecosystem biodiversity is the method of the agricultural management and land use. Agricultural systems that are used in modern agriculture may differently affect the environment, including biodiversity. Intensive agriculture is considered as the main reason of the decrease of flora and fauna species diversity and abundance in agroecosystems [14, 54]. The use of fertilizers and pesticides, removal of mid-field woody vegetation and bounds leading to fragmentation and degradations of habitats are among the most important threats of agricultural ecosystems [37]. Moreover, areas with worse conditions for agricultural production are abandoned or afforested.

Decreasing populations of the birds associated with the agricultural landscape in many European countries can serve as an example of the loss of biodiversity due to the intensification of methods of agricultural production and changes in the landscape [25]. Benton et al. [55] found a relationship between the changes in the population of birds associated with agricultural areas and the number of invertebrates and agricultural practices in Scotland. Intensive agriculture was also found to have a negative effect on other groups of organisms: soil microorganisms, weed flora, earthworms, insects, spiders, and mammals [19–20, 55–59]. The analyses performed by Storkey et al. [9] for 29 European countries showed a positive correlation between the yields of wheat and the number of endangered species. The study of the list of endangered or extinct species of wild plants in Germany showed that agriculture is responsible for the decrease of populations of 513 out of 711 species [19]. The endangered taxa included 10.8% of weeds. Fifteen species were considered extinct, which constituted 25% of all the extinct species. In Poland, about 60 percent of the 165 species of archeophytes that accompany crops are endangered, mainly due to the intensification of agriculture [60].

Species' ability to tolerate human impacts: destruction, degradation and fragmentation of habitats, reductions of individual survival and fecundity through exploitation, pollution and introduction of alien species varies among taxonomic groups [61]. For instance, the proportion of species listed as threatened in the International Union for Conservation of Nature Red List is much bigger in amphibians than in birds [62].

Intensification of agricultural practices causes the loss of biodiversity, and thus influence important ecosystem services. It affects plant production, plant protection, pollination, decomposition processes, nutrient cycles, and the resistance to invasive organisms [15, 63–65]. In some cases, the intensification of agricultural production can lead to an increase in the population of some, or even rare, species. A higher productivity of agricultural areas in comparison with natural ecosystems means more feed (biomass of plants and fruit) for birds, mammals, and butterflies [8]. Söderström et al. [66] found a greater abundance of bird species on the areas used for agriculture and the reduction of the diversity in the period after the abandonment of farming, while Westphal et al. [67] found an increase in the population of bumblebees together with the increase in the area of rape cultivation. Habitat value is, therefore, often determined by food resources, which result from high productivity, which in turn may have other negative environmental consequences.

Negative impacts of conventional farming on the environment, the overproduction of food, and consumer dissatisfaction with the quality of the products obtained through such farming, caused the development of the concept of sustainable agriculture, which uses environmentally friendly methods of production [68–69]. Such assumptions are the basis of the development of alternative systems of agricultural production, such as integrated and organic farming.

An integrated production system uses technical and biological progress in the cultivation, fertilization, and plant protection in a harmonious way, which allows to obtain a stable efficiency and a proper level of agricultural income through the use of methods that do not pose a threat to the environment. It combines the most important elements of organic and conventional farming, and allows for simultaneous realization of economic, ecological, and social goals [69]. Integrated production ensures sustainable economic development of the farm, takes into account the needs of the environment, and it is also attractive for consumers due to the obtained quality of products. The results of the implementation of the integrated system in several European countries show that it managed to significantly reduce the use of chemical pesticides and synthetic nitrogen fertilizers, which led to, among others, an increase in the diversity of flora and fauna [68, 70]. The Directive on the sustainable use of pesticides (2009/128/EC) [71] has obliged all EU member states to prepare and implement integrated crop protection programs, which to some extent can protect the biodiversity of flora and fauna [72].

One of the proposed solutions for combining productive and environmental functions of agriculture is an approach called "ecological intensification" [33]. For ecological intensification, the primary interest is in managing the processes and conditions that mediate yield levels. Ecological intensification entails the environmentally friendly replacement of anthropogenic inputs and/or enhancement of crop productivity, by including regulating and supporting ecosystem services management in agricultural practices. Research efforts and investments are particularly needed to reduce existing yield gaps by integrating context-appropriate bundles of ecosystem services into crop production systems.

6. The significance of biodiversity in organic farming

The aim of organic farming is the production of high-quality food and, at the same time, the protection of the environment [73–74]. The ecological system is fundamentally different from other systems of agricultural production because it excludes the use of synthetic mineral fertilizers, growth regulators, chemical plant protection products, and synthetic feed additives. It is based on substances of natural origin, which are not technologically processed [74]. Organic farming system is based on the use of environmentally friendly production methods that include crop rotations with a large share of legumes, organic fertilizers, and non-chemical methods of plant protection. Due to the resignation from the application of synthetic mineral fertilizers and chemical plant protection products, organic farming has an even greater positive impact on the diversity of flora and fauna than the integrated system [19, 22, 56, 59, 75–77]. The results of many studies point to the positive effects of organic farming on diversity of flora and fauna on arable lands and grasslands [76–81].

Dynamic development of organic farming is observed in the EU, including Poland [82]. Some authors believe that the dissemination of ecological system on agricultural areas may help reverse the negative trend of the decline of biodiversity in the cultivated fields, which was caused by the intensification of agriculture [19, 82].

The most direct way to capture the effects of human activities on biodiversity is to analyze time-series data from ecological communities or populations, relating changes in biodiversity to changes in human activities. Such long-term research (1996–2011) on weed flora diversity in different crop production systems, organic, integrated, and conventional, were conducted in the Experimental Station of the Institute of Soil Science and Plant Cultivation – State Research Institute (IUNG-PIB) in Puławy, Poland [N: 51º28', E: 22º04'] (Table 1).

Items	Crop production systems			
	Organic	Integrated	Conventional	Monoculture
Crop rotation	Potato Spring barley/spring wheat from 2005 + undersown crop Clovers and grasses (1st year) Clovers and grasses (2nd year) Winter wheat + catch crop	Potato Spring barley/spring wheat from 2005 + catch crop Faba bean or blue lupine Winter wheat + catch crop	Winter rape Winter wheat Spring barley/ spring wheat from 2005	Winter wheat
Seed dressing	-	+	+	
Organic fertilization	compost (30 t·ha⁻¹) under potato + catch crop	compost (30 t·ha⁻¹) under potato + 2 × catch crop	rape straw, winter wheat straw	wheat straw (every 2 years)
Mineral fertilization (kg·ha⁻¹)	according to the results of soil analysis, allowed P and K fertilizers in the form of natural rock	NPK (85+55+65)	NPK (140+60+80)	
Fungicide	-	2 x	2–3 x	
Retardants	-	1–2 x	2 x	
Weed control	weeder harrow 2–3 x	weeder harrow 1x herbicides 1–2 x	herbicides 2–3 x	

Table 1. Major elements of the agricultural practices of winter wheat in different farming systems (1996-2011); source [59].

The study showed that long-term management in organic system increased the diversity of weed flora accompanying crops (Figure 1). Simplifying the crop rotation from the integrated system, through the conventional system to monoculture of winter wheat, associated with the increased use of herbicides, led to the depletion of the species in weed communities. In the 16-year period, the average number of weed species in integrated and conventional systems, as well as in wheat monoculture was similar (6.1–6.8), while in the organic system by about 3.5 times higher (22 species). During the 16 years of research, the changes in weed communities in winter wheat cultivated in this farming system were found, especially involving the decreasing abundance of nitrophilous species: *Chenopodium album* and *Galium aparine* and the increasing density of more sensitive to herbicides taxa, *Stellaria media, Capsella bursa-pastoris, Fallopia convolvulus*, and species of the *Vicia* genus [59].

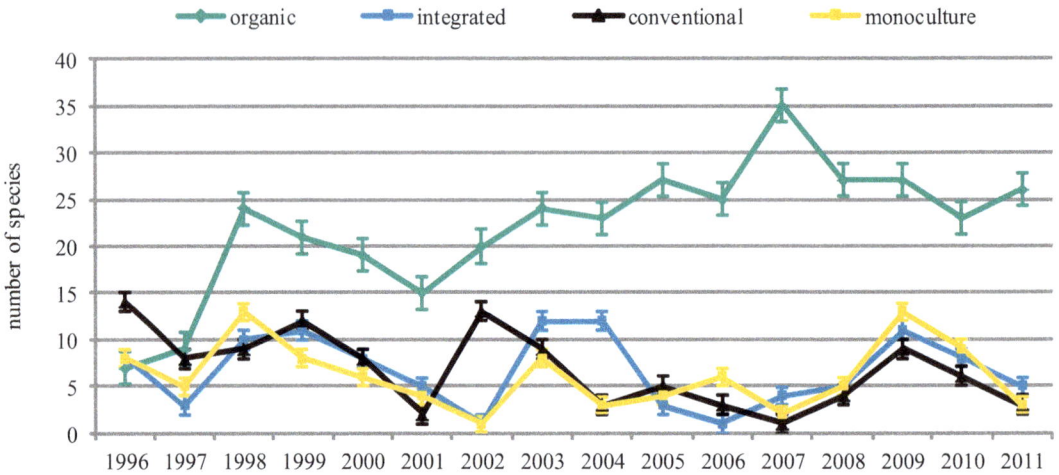

Figure 1. Weed plant diversity (± st. error) in winter wheat cultivated in different farming systems in years 1996–2011; source [59].

The agricultural practices applied in the compared farming systems (organic, integrated, conventional, and monoculture) of winter wheat differentiated the density of flora more than species composition. The largest number of weeds in the canopy of winter wheat at the dough stage was found in the organic system, 112 plants · m^{-2}, and the smallest for the integrated system, 18 plants · m^{-2}, on average (Figure 2). During the five years of the research (1997, 2001, 2002, 2007, 2008), the number of weeds in this treatment does not exceed 60 plants · m^{-2}, and only in two years (1996, 1999) was higher than 150 plants · m^{-2}, which means that it is possible to maintain weed infestation in organic cultivation of wheat at a relatively low level. Among the systems where herbicides were applied, the highest number of variability was observed in the monoculture of winter wheat.

Variability in species composition and abundance of weed flora throughout the years was influenced by the effectiveness of the applied methods of weed regulation and the weather conditions, which determined the germination of specific species of weeds and affected the density of wheat canopy and its competitiveness against weeds. In the systems where herbicides were applied, there were the highest fluctuations in the value of Shannon's and Simpson's indicators throughout the years (Figures 3 and 4). Shannon's diversity index value was the highest for weed flora in organic system and increased from 0.75 in 1996 to 2.64 in 2007 (Figure 3).

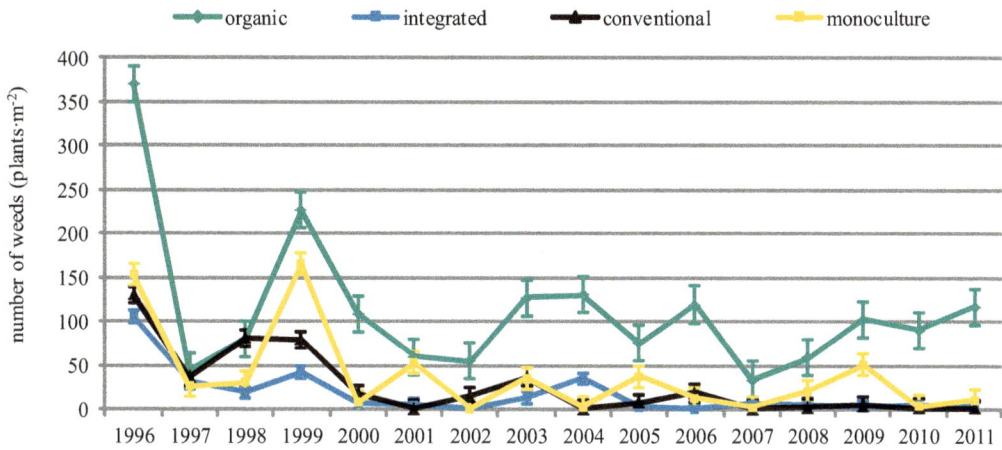

Figure 2. Weed abundance (± st. error) in winter wheat cultivated in different farming systems in years 1996–2011; source [59].

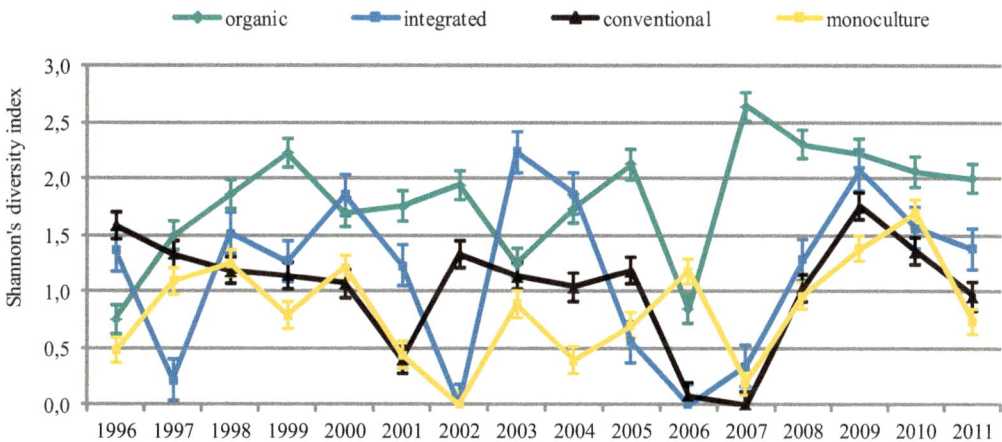

Figure 3. Shannon's diversity index values (± st. error) for weed communities in winter wheat cultivated in different farming systems in 1996–2011; source [59].

The dominance of some weed species in the community reflected in high Simpson's dominance index could affect the wheat yield more than diversified weed flora. A large diversity of weed species with low their quantity within species is less dangerous due to the yield because in multi-species weed community interspecies competition takes place. Interactions between weeds and the crop depend on the competitiveness and abundance of occurring weed species and the competitive abilities of the crop. In addition, those relationships are affected by environmental factors including soil conditions, weather, as well as agronomic practices.

It was found that weed communities in winter wheat cultivated in the organic system were characterized with a high qualitative and quantitative similarity in years, which was confirmed by the results of the ordination analysis (Figure 5).

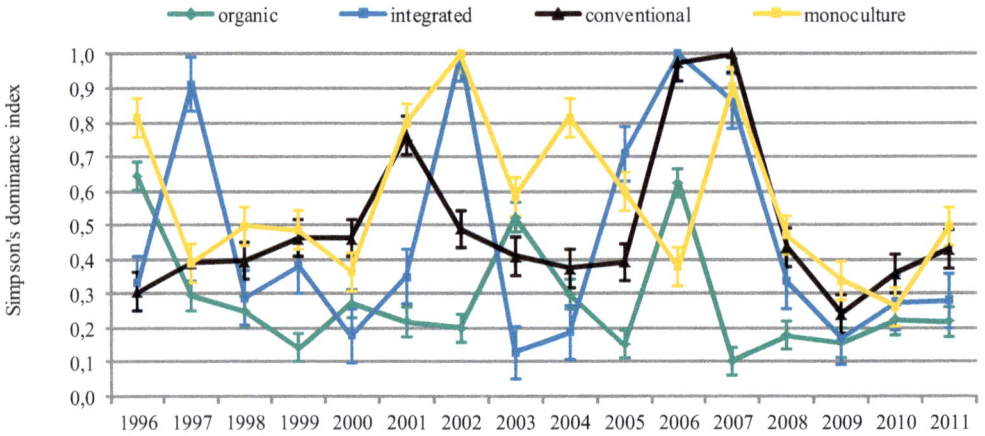

Figure 4. Simpson's dominance index values (± st. error) for weed communities in winter wheat cultivated in different farming systems in years 1996–2011; source [59].

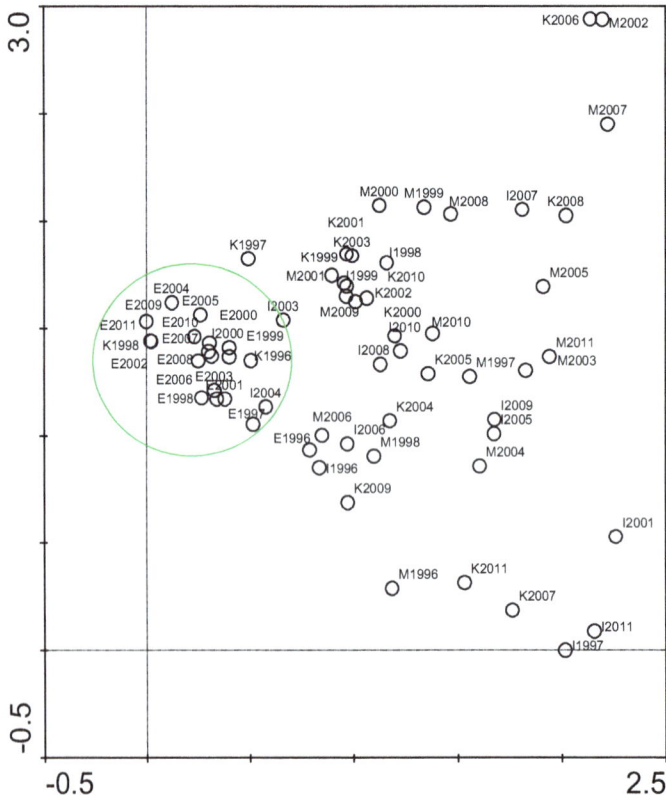

Figure 5. Ordination diagram of samples (represented weed flora communities in winter wheat cultivated in different crop production systems and years) in relation to first and second axes of Detrended Correspondence Analysis (DCA); source [59].

The comprehensive database that collates published, in-press, and other quality-assured spatial comparisons of community composition and site-level biodiversity from terrestrial sites around the world was created under the PREDICTS project (www.predicts.org.uk) [83]. Another example of a project that aimed to study the effect of different agricultural practices on diversity of flora, invertebrates, birds, and landscape in the east-south part of Poland and to prepare a geo-spatial database is the KIK/25 project (www. agropronatura.pl).

According to many research results, organic farming fulfills the promise to protect biodiversity better than conventional farming. Supporting farmers to convert their properties to organic land and to maintain organic farming within the scope of agri-environment schemes as a part of Common Agriculture Policy can have a significant impact in biodiversity as a result of management decisions farmers apply to their agricultural land [81].

7. Trends of changes in ecosystems and ecosystem services in the European Union

A large proportion of European biodiversity today depends on habitat provided by low-intensity farming practices, yet this resource is declining as European agriculture intensifies. Within the European Union, particularly the central and eastern new member states have retained relatively large areas of species-rich farmland; but despite increased investment in nature conservation here in recent years, farmland biodiversity trends appear to be worsening [11].

In the Report of the EU [84], analysis of the trends in the spatial extent of ecosystems and in the supply and use of ecosystem services at the European scale between 2000 and 2010 were presented. In the EU, urban land and forests increased while cropland, grassland, and heathland decreased (Figure 6). Many provisioning services showed increasing trends. Food and fodder crop production increased, even when agricultural areas decreased. More organic food was produced. More timber was removed from forests with increasing timber stocks. Total number of grazing livestock decreased.

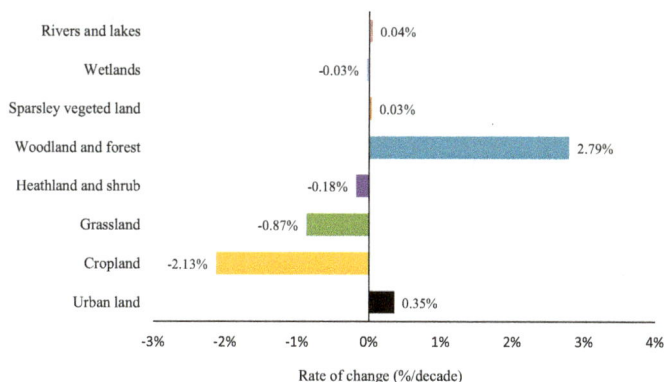

Figure 6. Change in the extent of surface area of ecosystems based on land cover data; source [84].

More area of natural environment was protected in 2010 than in 2000, but in contrast, the trends of two ecosystem services indicators that are directly related to biodiversity, pollination, and habitat quality were worsening (Figure 7). Crop production deficit was observed resulting from a loss of insect pollination. Habitat quality (regulation) slightly declined. There was a positive trend in the opportunity for citizens to have access to land with a high recreation potential.

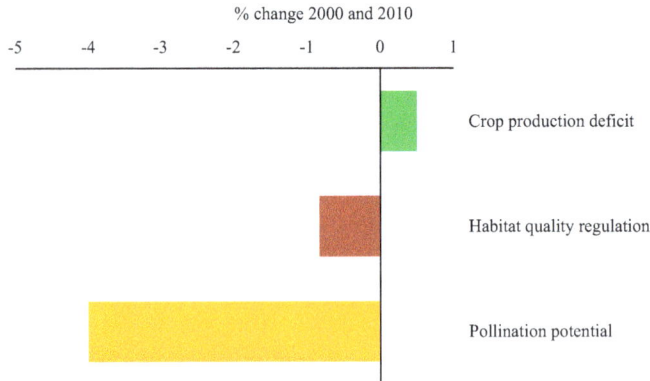

Figure 7. Main trends in ecosystem services in the EU between 2000 and 2010: Habitat maintenance and pollination; source [84].

Comparative studies show greater ecosystem quality for biodiversity as well as higher levels of rare species occurrence and species richness in lowland farmland in the central and eastern new member states than in Northern and Western Europe [11, 85]. In contrast to much of lowland EU, the main challenge and opportunity for farmland biodiversity conservation in the new member states is that a large number of species of conservation concern often still exist, e.g., in Polish field margins [11, 86]. These target species may have different requirements, creating conflicts when prescribing management measures. Simple but rigid measures applied over large areas can therefore be worse than existing management [11].

According to the EU Report, different trends in agriculture, ecosystems, and ecosystem services in EU countries were recorded (Figures 8 and 9) [84]. For example, in Poland relatively small changes were noted (increasing biomass built up and slightly negative trends in several services, including pollination potential) (Figure 8).

In France, where agriculture historically was more intensive than in Poland, slight decreases or status quo for many indicators were observed while the area under organic farming, timber stock, and forest area was rising (Figure 9).

Generally we see the following trends at the EU scale [84]:

For provisioning ecosystem services:

- More crops for food, feed, and energy are produced in the EU on less arable land. More organic food is grown. Textile crop production and the total number of grazing livestock have decreased.

POLAND

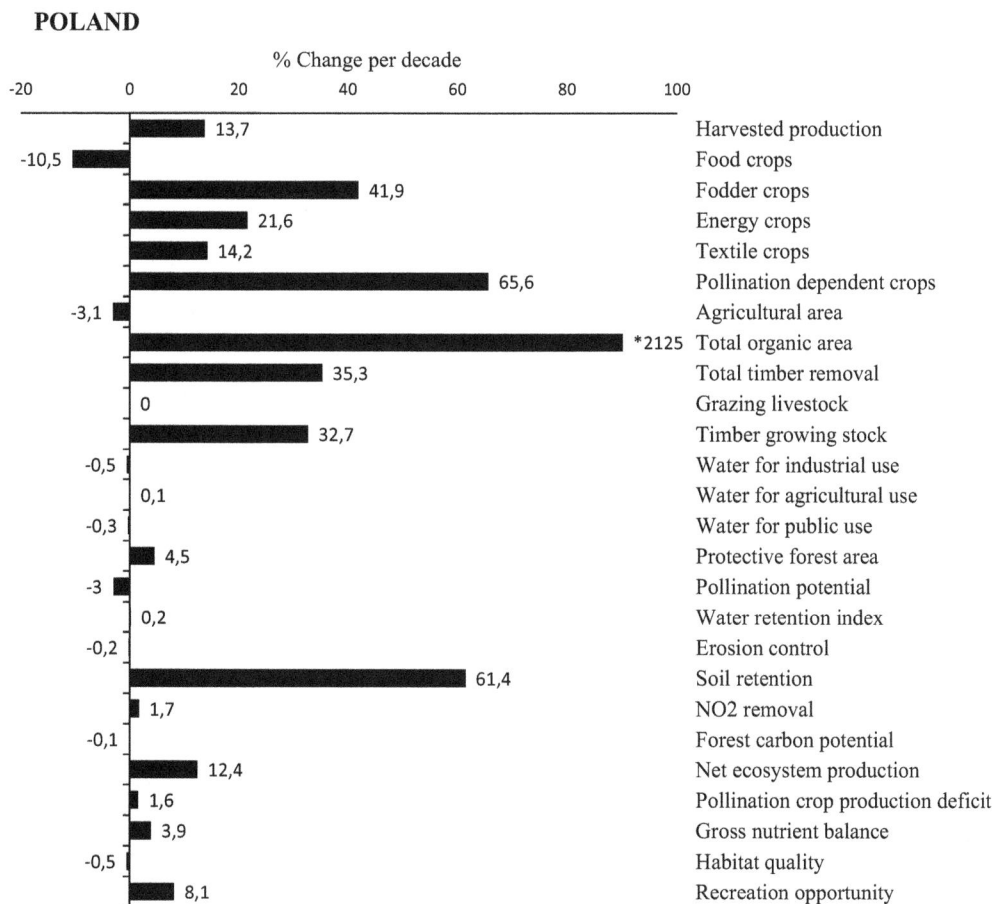

Figure 8. Trends in ecosystems and ecosystem services between 2000 and 2010 in Poland; source [84].

- The EU has used water in a slightly more resource-efficient way. Reported water abstractions decreased in both absolute and relative terms (relative to the naturally available water).

- Timber removals have increased and so, did the total timber stock.

For regulating ecosystem services:

- There is a substantial increase in net ecosystem productivity.

- Several regulating services, in particular those that are related to the presence of trees, woodland, or forests, increased slightly. This is the case for water retention, forest carbon potential, erosion control, and air quality regulation.

- Pollination potential and habitat quality show a negative trend.

For cultural ecosystem services:

- More land is protected and there is a positive trend in the opportunity for citizens to have access to land with a high recreation potential.

FRANCE

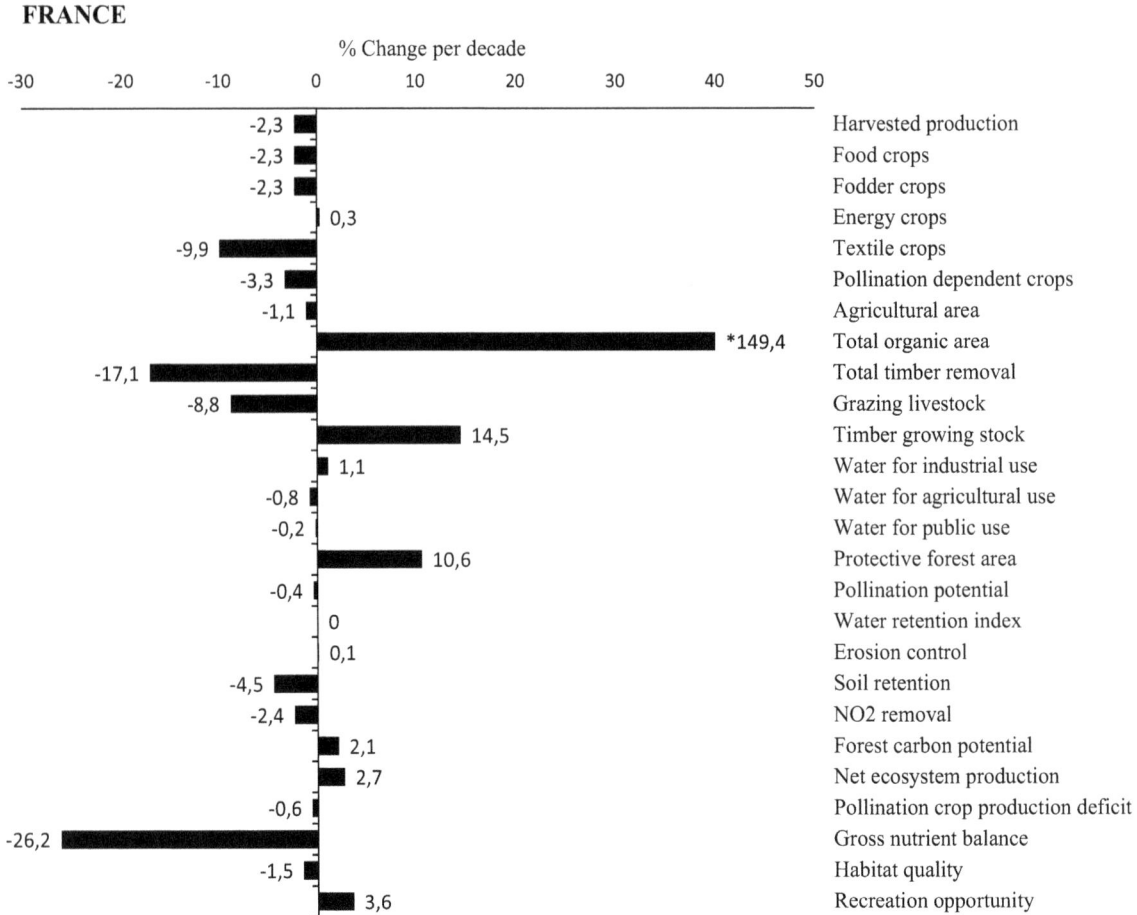

Figure 9. Trends in ecosystems and ecosystem services between 2000 and 2010 in France; source [84].

Costanza et al. [51] estimated the loss of global ecosystem services from 1997 to 2011 due to land use change at $4.3–20.2 billion/year, depending on which unit values were used. The biodiversity benefits for Europe and other countries of existing low-intensity farmland should be harnessed before they are lost. Instead of waiting for species-rich farmland to further decline, target research and monitoring to create locally appropriate conservation strategies for these habitats are needed now [11].

8. Summary

The protection of ecosystems and biodiversity is an important task and a key challenge to the world. The benefits of biodiversity conservation are difficult to notice in a short period of time or to economical evaluation. The benefits of the conservation of the species from extinction are important for future generations, because there may serve substances for medicine, genes useful in breeding, and others. At present, we do not know which plants may prove to be valuable in the future, which is why it is important to preserve as much gene pool as possible. Agriculture can contribute to the conservation of high-biodiversity systems, which may provide important ecosystem services such as pollination and biological control. Interdependencies between different groups of organisms, as well as the interaction between human activities and biodiversity require, however, further research. These studies should be conducted by experts from different disciplines in order to properly assess the value of biodiversity and ecosystem services, and create a strategy for the development of environmentally friendly agriculture and sustainable development of rural areas.

Acknowledgements

Publication was elaborated under the project "Protection of species diversity of valuable natural habitats on agricultural lands on Natura 2000 areas in the Lublin Voivodeship" (KIK/25) co-financed from the Swiss-Polish Cooperation Funds and multi-annual program of Institute of Soil Science and Plant Cultivation–State Research Institute, task 3.2. Assessment of the directions and agricultural production systems and the possibilities of their implementation in the regions and farms.

Author details

Beata Feledyn-Szewczyk*, Jan Kuś, Jarosław Stalenga, Adam K. Berbeć and Paweł Radzikowski

*Address all correspondence to: bszewczyk@iung.pulawy.pl

Institute of Soil Science and Plant Cultivation – State Research Institute, Department of Systems and Economics of Crop Production, Puławy, Poland

References

[1] United Nation 1992. Convention on Biological Diversity. United Nation Treaty Series. Rio de Janeiro. 5 June 1992; 1760, I-30619: 143-382.

[2] Clergue B, Amiaud B, Pervanchon F, Lasserre-Joulin F, Plantureux S. Biodiversity: Function and assessment in agricultural areas. A review. Agron. Sustain. Dev. 2005;25(1):1-15. DOI: 10.1051/agro:2004049.

[3] The economics of ecosystems and biodiversity. 2008. European Communities, Luksemburg, 64 p. ISBN-13 978-92-79-08960-2. http://ec.europa.eu/environment/nature/biodiversity/economics/pdf/teeb_report.pdf (Accessed 29 June 2015).

[4] Matson PA, Parton WJ, Power, Swift MJ. Agricultural intensification and ecosystem properties. Science 1997;277:504-509. DOI: 10.1126/science.277.5325.504.

[5] The UK National Ecosystem Assessment: Synthesis of the Key Findings. UNEP-WCMC, Cambridge, 2011; 85 p. ISBN: 978-92-807-3165-1.

[6] Environmental impact assessment of Rural Development Program project for 2007-2013. Warsaw, 2006.

[7] Report on the state of the environment in Poland 2008, Main Inspectorate for Environmental Protection, Environmental Monitoring Library, Warsaw 2010, 123 p.

[8] Tscharntke T, Klein AM, Kruess A, Steffan-Dewenter I, Thies C. Landscape perspectives on agricultural intensification and biodiversity—ecosystem service management. Ecol. Lett. 2005;8:857-874. DOI: 10.1111/j.1461-0248.2005.00782.x.

[9] Storkey J, Meyer S, Still KS, Leuschner C. The impact of agricultural intensification and land-use change on the European arable flora. Proc. R. Soc. B. 2012;279:1421-1429. DOI: 10.1098/rspb.2011.1686.

[10] EEA 2011. Annual report 2010 and Environmental statement 2011, 2011; 95 p. DOI: 10.2800/72655.

[11] Sutcliffe LME, Batary P, Kormann U, et al. Harnessing the biodiversity value of Central and Eastern European farmland. Diversity Distrib. 2015;21:722-730. DOI: 10.1111/ddi.12288.

[12] Mace GM, Cramer W, Diaz S, Faith DP, Larigauderie A, Le Prestre P, et al. Biodiversity targets after 2010. Curr. Opin. Environ. Sustain. 2010;2:3-8.

[13] European Commission. The EU Biodiversity Strategy to 2020. 2011. http://eur-lex.europa.eu/legal-content/EN/TXT/PDF/?uri=CELEX:52011DC0244&from=EN (Accessed 29 June 2015).

[14] Tryjanowski P, Dajok Z, Kujawa K, Kałuski T, Mrówczyński M. Threats to biodiversity in farmland: Are results from Western Europe good solution for Poland? Polish J. Agron. 2011;7:113-119.

[15] Rosin ZM, Takacs V, Báldi A, Banaszak-Cibicka W, Dajdok Z, Dolata PT, Kwieciński Z, Łangowska A, Moroń D, Skórka P, Tobółka M, Tryjanowski P, Wuczyński A. Ecosystem services as an efficient tool of nature conservation: A view from the Polish farmland. Chrońmy Przyr. Ojcz. 2011;67(1):3-20.

[16] Food and Agriculture Organization of the United Nations. What is agrobiodiversity? In: Building on Gender, Agrobiodiversity and Local Knowledge, 2004.

[17] Agri-environmental program guide 2007-2013. Ministry of Agriculture and Rural Development, Warsaw, 2009, 32 p.

[18] Thill D C, Lish J M, Callihan R H, Bechinski E J. Integrated weed management—a component of integrated pest management: a critical review. Weed Technol. 1991;5:648-656.

[19] Van Elsen T. Species diversity as a task for organic agriculture in Europe. Agric. Ecosyst. Environ. 2000;77:101-109.

[20] Marshall EJP, Brown VK, Boatman ND, Lutman PJW, Squire GR, Ward LK. The role of weeds in supporting biological diversity within crop fields. Weed Res. 2003; 43(2): 77-89. DOI: 10.1046/j.1365-3180.2003.00326.x.

[21] Hochół T. Weeds or plants accompanying crops. Pamiętnik Puławski 2003;134:90-96.

[22] Hole D G, Perkins A J, Wilson J D, Alexander I H, Grice P V, Evans A D. Does organic farming benefit biodiversity? Biol. Conserv. 2005;122:113-130. DOI:10.1016/ j.biocon.2004.07.018.

[23] Gerowitt B, Bertke E, Hespelt SK, Tute C. Towards multifunctional agriculture– weeds as ecological goods? Weed Res. 2003;43:227-235. DOI: 10.1046/j. 1365-3180.2003.00340.x.

[24] Wilson PJ, Aebischer NJ. The distribution of dicotyledonous weeds in relation to distance from the field edge. J. Appl. Ecol. 1995;32:295-310. DOI: 10.2307/2405097.

[25] Chamberlain DE, Fuller RJ, Bunce RGH, Duckworth JC, Shrubb M. Changes in the abundance of farmland birds in relation to the timing of agricultural intensification in England and Wales. J. Appl. Ecol. 2000;37:771-188. DOI: 10.1046/j. 1365-2664.2000.00548.x.

[26] Wilson JD, Morris AJ, Arroyo BE, Clark SC, Bradbury RB. A review of the abundance and diversity of invertebrate and plant foods of granivorous birds in northern Europe in relation to agricultural change. Agric. Ecosyst. Environ. 1999;75:13-30.

[27] Maloney D, Drummond FA, Alford R. Spider predation in agroecosystems: Can spiders effectively control pest population? Maine Agricultural and Forest Experiments Station Technical Bulletin. 2003;340:190.

[28] Sunderland K. Mechanisms underlying the effects of spiders on pest populations. Journal of Arachnology. 1999:308-316.

[29] Twardowski JP, Pastuszko K. Coastal habitats in agrocenosis of winter wheat as reservoirs of beneficial beetles (Col. Carabidae). J. Res. Appl. Agric. Eng. 2008;53(4): 123-127.

[30] Purtauf T, Roschewitz I, Dauber J, Thies C, Tscharntke T, Wolters V. Landscape context of organic and conventional farms: Influences on carabid beetle diversity. Agriculture, Ecosystems & Environment. 2005;108(2):165-174. DOI:10.1016/j.agee. 2005.01.005.

[31] Snyder WE, Ives AR. Interactions between specialist and generalist natural enemies: Parasitoids, predators, and pea aphid biocontrol. Ecology 2003;84(1):91-107.

[32] Flohre A, Rudnick M, Traser G, Tscharntke T, Eggers T. Does soil biota benefit from organic farming in complex vs. simple landscape? Agr. Ecosyst. Environ. 2011;141(1-2):210-214. DOI: 10.1016/j.agee.2011.02.032.

[33] Bommarco R, Kleijn D, Potts SG. Ecological intensification: Harnessing ecosystem services or food security. Trends in Ecology and Evolution. 2013;28:230-238. DOI: 10.1016/j.tree.2012.10.012.

[34] Costanza R, D'Arge R, De Groot R, Farberk S, Grasso M, Bruce Hannon B, Limburg K, Naeem S, O'Neill R V, Paruelo J, Raskin R G, Suttonkk P, van den Belt M. The value of the world's ecosystem services and natural capital. Nature. 1997;387:253–260. DOI:10.1038/387253a0.

[35] MEA 2005. The Millenium Ecosystem Assessment, Ecosystems and Human Well-being: Synthesis, Island Press, Washington, DC.

[36] Hyvönen T, Huusela-Veistola E. Arable weeds as indicators of agricultural intensity —a case study from Finland. Biol. Conserv. 2008;141:2857-2864.

[37] Mace G M, Norris K, Fitter A H. Biodiversity and ecosystem services: A multilayered relationship. Trends in Ecology and Evolution. 2012;27(1):19-26. DOI: 10.1016/j.tree.2011.08.006.

[38] Fischer A, Young J C. Understanding mental constructs of biodiversity: Implications for biodiversity management and conservation. Biol. Conserv. 2007;136:271-282. DOI: 10.1016/j.biocon.2006.11.024.

[39] Balvanera P, Pfisterer Andrea B, Buchmann N, He Jing-Shen, Nakashizuka T, Raffaelli D, Schmid B. Quantifying the evidence for biodiversity effects on ecosystem functioning and sevices, Ecol. Lett. 2006;9:1146-1156. DOI: 10.1111/j.1461-0248.2006.00963.x.

[40] Klein A-M, Vaissiere BE, Cane JH, Steffan-Dewenter I, Cunningham SA, Kremen C, Tscharntke T. Importance of pollinators in changing landscapes for world crops. Proceedings of the Royal Society of London. Series B, Biological Sciences. 2007;274(1608): 303-313. DOI:10.1098/rspb.2006.3721.

[41] Gallai N, Salles JM, Settele J, Vaissičre BE. Economic valuation of the vulnerability of world agriculture confronted with pollinator decline. Ecol. Econ. 2008;68:810-821.

[42] Kremen C, Williams NM, Bugg RL, Fay JP, Thorp RW. The area requirement of an ecosystem service: Crop pollination by native bee communities in California. Ecol. Lett. 2004;7:1109-1119. DOI: 10.1111/j.1461-0248.2004.00662.x.

[43] Loreau M, Naeem S, Inchausti P, Bengtsson J, Grime JP, Hector A, Hooper DU, Huston MA, Rafaelli D, Schmid B, Timan D, Wardle DA. Biodiversity and ecosystem

functioning: Current knowledge and future challenges. Science. 2001;294:804-808. DOI:10.1126/science.1064088.

[44] Costanza R, Fisher B, Mulder K, Liu S, Christopher T. Biodiversity and ecosystem services: A multi-scale empirical study of the relationship between species richness and net primary production. Ecological Economics. 2007;61:478-491.

[45] Bullock JM, Pywell RF, Burke MJW, Walker KJ. Restoration of biodiversity enhances agricultural production. Ecol. Lett. 2001;4:185-189.

[46] Lavelle P T, Decaens M, Aubert M, Barot S, Blouin M, Bureau F, Margerie P, Mora P, Rossi J-P. Soil invertebrates and ecosystem services. European Journal of Soil Biology. 2006; 42(Supplement 1): S3-S15.

[47] Hillebrandt H, Matthiessen B. Biodiversity in a complex world: Consolidation and progress in functional biodiversity research. Ecol. Lett. 2009;12:1405-1419. DOI: 10.1111/j.1461-0248.2009.01388.x.

[48] Harrison PA, et al. Linkages between biodiversity attributes and ecosystem services: A systematic review. Ecosystem Services. 2014;9:191. DOI: 10.1016/j.ecoser. 2014.05.006.

[49] Grime JP. Biodiversity and ecosystem function: The debate deepens. Science. 1997;277: 1260-1261. DOI: 10.2307/2892490.

[50] Rodriguez MA, Hawkins BA. Diversity, function and stability in parasitoid communities. Ecol. Lett. 2000;3:35-40.

[51] Costanza R, de Groot R, Sutton P, van der Ploeg S, Anderson S J, Kubiszewski I, Farber S, Turner R K. Changes in the global value of ecosystem services. Global Environmental Change. 2014;26:152-158. DOI: 10.1016/j.gloenvcha.2014.04.002.

[52] Solon J. "Ecosystem Services" concept and its application in landscape-ecological studies. In: Chmielewski TJ (ed.) The structure and functioning of the landscape: A meta-analysis, models, theories, and their applications. Problemy Ekologii Krajobrazu. 2008;21:25-44.

[53] de Groot RS, Wilson MA, Boumans RMJ. A typology for the classification, description and valuation of ecosystem functions, goods and services. Ecological Economics. 2002;41:393-408.

[54] Andreasen C, Stryhn H, Streibig JC. Decline of the flora in Danish arable fields. J. Appl. Ecol. 1996;33:619-626.

[55] Benton TG, Bryant DM, Cole L, Crick HQP. Linking agricultural practice to insect and bird populations: A historical study over three decades. J. Appl. Ecol. 2002;39:273-287.

[56] Hyvönen T, Ketoja E, Salonen J, Jalli H, Tiainen J. Weed species diversity and community composition in organic and conventional cropping of spring cereals. Agric. Ecosyst. Environ. 2003;97:131-149.

[57] Irmler U. Changes in earthworm populations during conversion from conventional to organic farming. Agric. Ecosyst. Environ. 2010;135(3):194-198.

[58] Flohre A, Rudnick M, Traser G, Tscharntke T, Eggers T. Does soil biota benefit from organic farming in complex vs. simple landscape? Agr. Ecosyst. Environ. 2011;141(1-2):210-214. DOI:10.1016/j.agee.2011.02.032.

[59] Feledyn-Szewczyk B. The influence of agricultural land use on weed flora diversity. Monografie i Rozprawy Naukowe IUNG-PIB. 2013;36:184.

[60] Zając M, Zając A, Tokarska-Guzik B. Extinct and endangered archaeophytes and the dynamics of their diversity in Poland. Biodiv. Res. Conserv. 2009;13:17-24. DOI: 10.2478/v10119-009-0004-4.

[61] Vié J, Hilton-Taylor C, Stuart SN. Wildlife in a Changing World: An analysis of the 2008 IUCN Red List of Threatened Species. International Union for Conservation of Nature, Gland, Swizerland, 2009, 180 p.

[62] International Union for Conservation of Nature. 2013. The IUCN red list of threatened species. http://www.iucnredlist.org/ (Accessed 29 June 2015).

[63] Donald PF. Biodiversity impacts of some agricultural commodity production systems. Conserv. Biol. 2004;18:17-37.

[64] Klein A-M, Vaissiere BE, Cane JH, Steffan-Dewenter I, Cunningham SA, Kremen C, Tscharntke T. Importance of pollinators in changing landscapes for world crops. Proceedings of the Royal Society of London. Series B, Biological Sciences, 2007;274(1608): 303-313.

[65] Kennedy TA, Naeem S, Howe KM, Knops JMH, Tilman D, Reich P. Biodiversity as a barrier to ecological invasion. Nature. 2002;417:636-638.

[66] Söderström B, Svensson B, Vessby K, Glimskar A. Plants, insects and birds in semi-natural pastures in relation to local habitat and landscape factors. Biodivers. Conserv. 2001;10:1839-1863.

[67] Westphal C, Steffan-Dewenter I, Tscharntke T. Mass-flowering crops enhance pollinator densities at a landscape scale. Ecol. Let. 2003;6:961-965.

[68] Jordan VWL. Opportunities and constrains for integrated farming system. Proc. 2nd ESA Congress, Warwick Univ., 1992, pp. 318-325.

[69] Kuś J. Farming systems. Integrated farming. IUNG. 1995;42/95:28.

[70] Kuś J, Stalenga J. Prospects for the development of different agricultural systems of production in Poland. Biul. IHAR. 2006;242:15-25.

[71] Directive 2009/128/EC of the European Parliament and the Council of 21 October 2009 establishing a framework for Community action to achieve the sustainable use of pesticides. http://eur-lex.europa.eu/legal-content/EN/TXT/PDF/?uri=CELEX: 32009L0128&from=EN (Acessed 29 June 2015).

[72] Pruszyński S. Plant protection in different production systems and biodiversity. Prog. Plant Protect. 2009;49(3):1091-1101.

[73] Council Regulation (EC) No 834/2007 of 28 June 2007 on organic production and labelling of organic products and repealing Regulation (EEC) No 2092/91. http://eur-lex.europa.eu/LexUriServ/LexUriServ.do?uri=OJ:L:2007:189:0001:0023:EN:PDF (Accessed 29 June 2015).

[74] Kuś J. Farming systems. Organic farming. IUNG, 1995;45/95:62.

[75] Hald AB. Weed vegetation (wild flora) of long established organic versus conventional cereal fields in Denmark. Ann. Appl. Biol. 1999;14:307-314.

[76] Pfiffner L, Luka H. Effects of low-input farming systems on carabids and epigeial spiders—a paired farm approach. Basic Appl. Ecol. 2003;4:117-127.

[77] Bengtsson J, Ahnström J, Weibull AC. The effects of organic agriculture on biodiversity and abundance: A meta-analysis. J. Appl. Ecol. 2005;42:261-269.

[78] Moreby SJ, Aebischer NJ, Souhway SE, Sotherton NW. A comparison of the flora and arthropod fauna of organically and conventionally grown winter wheat in southern England. Ann. Appl. Biol. 1994;125:13-27.

[79] Stolze M, Piorr A, Häring A, Dabbert S. The Environmental Impacts of Organic Farming in Europe. Organic Farming in Europe: Economics and Policy, Stuttgart, University of Hohenheim, 2000;6:23-90.

[80] Fuller RJ, Norton LR, Feber RE, Johnson PJ, Chamberlain DE, Joys AC, Mathews F, Stuart RC, Townsend MC, Manley WJ, Wolfe MS, Macdonald DW, Firbank LG. Benefits of organic farming to biodiversity vary among taxa. Biol. Lett. 2005;1:431-434.

[81] Bavec M, Bavec F. Impact of Organic Farming on Biodiversity. In: Biodiversity in Ecosystems—Linking Structure and Function, Yueh-Hsin Lo, Juan A. Blanco and Shovonlal Roy (eds.), 2015;8:185-202. DOI: 10.5772/58974.

[82] Hyvönen T. Can conversion to organic farming restore the species composition of arable weed communities? Biol. Conserv. 2007;137:382-390.

[83] Hudson LN, Newbold T, Contu S. et. al. The PREDICTS database: A global database of how local terrestrial biodiversity responds to human impacts. Ecology and Evolution. 2014;1-35. DOI: 10.1002/ece3.1303.

[84] Mapping and Assessment of Ecosystems and their Services: Trends in ecosystems and ecosystem services in the European Union between 2000 and 2010. European Commission – Joint Research Centre – Institute for Environment and Sustainability,

Luxembourg: Publications Office of the European Union, 2015;131. DOI: 10.2788/341839.

[85] Batáry P, Báldi A, Sárospataki M, Kohler F, Verhulst J, Knop E, Herzog F, Kleijn D. Effect of conservation management on bees and insect-pollinated grassland plant communites in three European countries. Agriculture, Ecosystem and Environment. 2010;136:35-39.

[86] Wuczyński A, Dajdok Z, Wierzcholska S, Kujawa K. Applying red list to the evaluation of agricultural habitat: Regular occurrence of threatened birds, vascular plants, and bryophytes in field margins of Poland. Biodiversity and Conservation. 2014;23:999-1017.

Permissions

All chapters in this book were first published in OF, by InTech Open; hereby published with permission under the Creative Commons Attribution License or equivalent. Every chapter published in this book has been scrutinized by our experts. Their significance has been extensively debated. The topics covered herein carry significant findings which will fuel the growth of the discipline. They may even be implemented as practical applications or may be referred to as a beginning point for another development.

The contributors of this book come from diverse backgrounds, making this book a truly international effort. This book will bring forth new frontiers with its revolutionizing research information and detailed analysis of the nascent developments around the world.

We would like to thank all the contributing authors for lending their expertise to make the book truly unique. They have played a crucial role in the development of this book. Without their invaluable contributions this book wouldn't have been possible. They have made vital efforts to compile up to date information on the varied aspects of this subject to make this book a valuable addition to the collection of many professionals and students.

This book was conceptualized with the vision of imparting up-to-date information and advanced data in this field. To ensure the same, a matchless editorial board was set up. Every individual on the board went through rigorous rounds of assessment to prove their worth. After which they invested a large part of their time researching and compiling the most relevant data for our readers.

The editorial board has been involved in producing this book since its inception. They have spent rigorous hours researching and exploring the diverse topics which have resulted in the successful publishing of this book. They have passed on their knowledge of decades through this book. To expedite this challenging task, the publisher supported the team at every step. A small team of assistant editors was also appointed to further simplify the editing procedure and attain best results for the readers.

Apart from the editorial board, the designing team has also invested a significant amount of their time in understanding the subject and creating the most relevant covers. They scrutinized every image to scout for the most suitable representation of the subject and create an appropriate cover for the book.

The publishing team has been an ardent support to the editorial, designing and production team. Their endless efforts to recruit the best for this project, has resulted in the accomplishment of this book. They are a veteran in the field of academics and their pool of knowledge is as vast as their experience in printing. Their expertise and guidance has proved useful at every step. Their uncompromising quality standards have made this book an exceptional effort. Their encouragement from time to time has been an inspiration for everyone.

The publisher and the editorial board hope that this book will prove to be a valuable piece of knowledge for researchers, students, practitioners and scholars across the globe.

List of Contributors

Sanjay Kumar Yadav, Subhash Babu, Gulab Singh Yadav, Raghavendra Singh and Manoj Kumar Yadav
ICAR-Central Potato Research Station, Upper Shillong, Meghalaya, India

Andrew J. Price, Leah M. Duzy, Kip S. Balkcom and Ted S. Kornecki
United States Department of Agriculture, Agricultural Research Service, National Soil Dynamics Laboratory, Auburn, Alabama, USA

Jessica A. Kelton
Auburn University, Auburn, Alabama, USA

Lina Sarunaite
Institute of Agriculture, Lithuanian Research Centre for Agriculture and Forestry, Lithuania

Juan C. Angeles Hernandez
Programa de Maestría y Doctorado en Ciencias de la Producción y de la Salud Animal, Facultad de Medicina Veterinaria y Zootecnia, Universidad Nacional Autónoma de México, Circuito Exterior, Ciudad Universitaria, Delegación Coyoacán, México Facultad de Medicina Veterinaria y Zootecnia, Departamento de Nutrición Animal y Bioquímica, Universidad Nacional Autónoma de México, Circuito Exterior, Ciudad Universitaria, Delegación Coyoacán, México

Octavio A. Castelan Ortega
Universidad Autónoma del Estado de México, Instituto Literario 100 Ote, Toluca, Estado de México, México

Sergio Radic Schilling
Escuela de Ciencias y Tecnologías en Recursos Agrícolas y Acuícolas, Universidad de Magallanes, Región de Magallanes y la Antártica Chilena, Chile

Sergio Angeles Campos and A. Hilda Ramirez Perez
Facultad de Medicina Veterinaria y Zootecnia, Departamento de Nutrición Animal y Bioquímica, Universidad Nacional Autónoma de México, Circuito Exterior, Ciudad

Manuel Gonzalez Ronquillo
Universidad Autónoma del Estado de México, Instituto Literario 100 Ote, Toluca, Estado de México, México

Suarau O. Oshunsanya and OrevaOghene Aliku
Department of Agronomy, University of Ibadan, Ibadan, Nigeria

Mônica Mateus Florião
Rural Federal University of Rio de Janeiro - UFRRJ, Brazil

Wagner Tassinari
Department of Mathematics, Institute of Mathematical Sciences, UFRRJ, Rio de Janeiro, Brazil

Suja Girija, Sreekumar Janardanan, Jyothi Alummoottil Narayanan and Santosh Mithra Velayudhan Santhakumari
ICAR-Central Tuber Crops Research Institute, Thiruvananthapuram, Kerala, India

Petr Dvořák, Jaroslav Tomášek, Karel Hamouz and Michaela Jedličková
Czech University of Life Sciences Prague, Faculty of Agrobiology, Food and Natural Resources, Kamýcká, Prague – Suchdol, Czech Republic

Beata Feledyn-Szewczyk, Jan Kuś, Jarosław Stalenga, Adam K. Berbeć and Paweł Radzikowski
Institute of Soil Science and Plant Cultivation – State Research Institute, Department of Systems and Economics of Crop Production, Puławy, Poland Universitaria, Delegación Coyoacán, México

Index